We Can Do It!

Year 4

USING AND APPLYING MATHS CHALLENGES

Peter Clarke

Acknowledgement

The author wishes to thank Mike Askew, Sheila Ebbutt and Brian Molyneaux for their valuable contributions to this publication.

Thanks also to the following:

Karen Holman, Paddox Primary School, Rugby
Hilary Head, Send C of E First School, Surrey
Shirley Mulroy, Leslie Rankin, Hady Primary School, Chesterfield
Catherine Aket, Whitecrest Primary School, West Midlands
Kim Varden, Broke Hall Primary School, Suffolk
Elaine Richardson, St Augustine's Primary School, Cheshire
Carolyn Wallis, St Nicholas House Junior School, Hertfordshire
Lyn Wickham, Emma Bailey, Bidbury Junior School, Hampshire
Mandy Patterson, Temple Mill Primary School, Kent
John Ellard, Kingsley Primary School, Northampton
Father Rudolf Loewenstein, St Christina's Primary School, Camden
Joyce Lydford, Balgowan Primary School, Kent
Louise Guthrie, Angela Beall, Bardsey Primary School, Leeds
Sharon Thomas, Cwmbwrla Primary School, Swansea
Lynwen Barnsley, Education Effectiveness, Swansea
Jennie Jump, Advisor, Leeds
Sharon Sutton, University of Reading
Steve Lumb, Fielding Primary School, Ealing
Deborah de Gray, West Kingsdown C of E Primary School, Kent
Kerry Ann Darlington, Ullapool Primary School, Ross-shire
Helen Elis Jones, University of Wales, Bangor
Jayne Featherstone, Elton Community Primary School, Lancashire
Jane Airey, Frith Manor Primary School, Barnet
Andrea Trigg, Felbridge Primary School, West Sussex
Helen Andrews, Blue Coat School, Birmingham
Joyce Atkinson, Croham Hurst Junior School, Surrey
Jane Holmes, Elizabeth Wyles, St John's Primary School, Oxon

Thanks also to the BEAM Development Group:

Mich Bahn, Canonbury Primary School, Islington
Joanne Barrett, Rotherfield Primary School, Islington
Catherine Horton, St Jude and St Paul's School, Islington
Simone de Juan, Prior Weston Primary School, Islington

Published by BEAM Education
Maze Workshops
72a Southgate Road
London N1 3JT

Telephone 020 7684 3323
Fax 020 7684 3334
Email info@beam.co.uk
www.beam.co.uk

ISBN 978 1 9062 2448 6
British Library Cataloguing-in-Publication Data
Data available
Edited by Marion Dill
Design by Reena Kataria
Layout by Matt Carr and Redmoor Design
Illustrations by Annabel Hudson and Matt Carr
Cover photo: Lauriston School, Hackney
Printed by Graphy Cems, Spain

Contents

Introduction

What is AT1: Using and applying mathematics?

Mathematical problem solving involves using previously acquired mathematical understanding, knowledge and skills and applying these to solve problems arising within everyday life, as well as within mathematics.

A major reason for studying mathematics is to be able to develop problem-solving, reasoning and logical skills that we can apply to everyday situations. We want children to employ their 'pure' mathematical knowledge effectively in real-life situations day by day, both within and outside school. It is important for children to see how acquiring mathematical understanding can help them solve problems that are relevant to their daily lives. Mathematics can help them interpret and analyse real-life situations. It can also be a source of creative pleasure for its own sake.

Using and applying mathematics is often mistaken as simply solving word problems.

Tomas shared 20 marbles equally among himself and his four friends. How many marbles did each child get?

To solve this word problem, you need to be able to read and understand what the problem is, identify the calculation you need to do, do the calculation and then interpret the answer you get in the context of the problem. In a limited sense, you are using and applying mathematics, but at a low level. Word problems at this level follow a predictable pattern, which removes the need for any real problem solving.

Of course, word problems can be more complex, and children have to work out which information is relevant, and what the context tells you about the kind of answer you need (and knowing key words is not always a helpful strategy). This multi-step problem is more like the kinds of problems that children face in real life.

Tomas is 8 years old, and he gets 24 marbles. He keeps half of the marbles for himself and shares out the rest equally among his four friends. How many more marbles does he have than each of his friends?

The point about more complex problems is that you have to work out what the meaning is, what sort of outcome you need, and what sensible calculations to do to get there. Problems in real life are mainly like this.

An investigative approach to the simple marble problem could be:

How many different ways can you share 20 marbles equally?

A more complex investigation could be:

It's Tomas's birthday, so he gets more marbles than anyone else. Everyone else gets the same amount. How many different ways can you share the 24 marbles?

Sometimes it is interesting just exploring some mathematics for its own sake, as a pastime. Some of the problems in ***We Can Do It!*** are like that. The interest lies in using your brain, finding a pattern, seeking a neat solution – just like doing a crossword or sudoku puzzle. In the process of working on the investigation, children will be honing their reasoning skills and using their creativity to seek a way forward.

Some problems really do mirror everyday life:

If we spend less time tidying up, will we get more time for playing outside?

What news stories on the front page of the newspaper are given the most space?

What proportion of water do you need to dilute a fruit drink?

Many of these involve measures and data handling. Children need practical experiences to solve these: if you don't know what 100 ml of liquid looks like, and how it compares with 1 litre, you cannot solve the problem in the abstract. This means being prepared for a busy, active and, at times, messy classroom – and also a classroom where children discuss with each other the problem in hand.

Real and complex problems and investigations require children to search for strategies to get started and to draw upon their experiences and knowledge of 'pure' mathematics. They also encourage children to work flexibly, creatively and logically. They are less comfortable for the teacher because the outcomes are not always predictable and the answers are not always known. Our role is to work with children, sometimes doing the mathematics alongside them, looking for and encouraging creative and logical thinking, rather than focusing on right answers.

Mathematical thinking

The *National Curriculum* (2000) outlines the thinking skills that complement the key understanding, knowledge and skills that are embedded in the statutory primary curriculum.

The ***We Can Do It!*** series aims to develop the following key thinking skills in children:

Information – processing skills

- Locate, collect relevant information
- Sort, classify, sequence, compare and analyse part and/or whole relationships

Reasoning skills

- Give reasons for opinions and actions
- Draw inferences and make deductions
- Use precise language to explain what they think
- Make judgements and decisions informed by reason or evidence

Enquiry skills

- Ask relevant questions
- Pose and define problems
- Plan what to do and how to research
- Predict outcomes and anticipate conclusions
- Test conclusions and improve ideas

Creative thinking skills

- Generate and extend ideas
- Suggest hypotheses
- Apply imagination
- Look for alternative innovative outcomes

Evaluative skills

- Evaluate information
- Judge the value of what they read, hear or do
- Develop criteria for judging the value of their own and others' work or ideas
- Have confidence in their judgement

We Can Do It!

In this series, we provide problems and challenges that stimulate genuine mathematical thinking. These problems are written for a community of learners in the primary classroom – that is, we expect the problems to be solved collaboratively by pairs of children, groups and whole classes working together and discussing the problems at every stage. With each problem, we offer teaching advice on how to encourage high-level thinking among children. We also analyse each problem and children's possible responses to it in order to promote greater understanding of how children develop problem-solving skills.

The challenges in ***We Can Do It!*** are designed to improve children's attainment in the three strands of AT1 of the *National Curriculum* (2000): Using and applying mathematics.

In ***problem solving*** by:

- using a range of problem solving strategies
- trying different approaches to a problem
- applying mathematics in a new context
- checking their results

In ***communicating*** by:

- interpreting information
- recording information systematically
- using mathematical language, symbols, notation and diagrams correctly and precisely
- presenting and interpreting methods, solutions and conclusions in the context of the problem

In ***reasoning*** by:

- giving clear explanations of their methods and reasoning
- investigating and making general statements
- recognising patterns in their results
- making use of a wider range of evidence to justify results through logical reasoned argument
- drawing their own conclusions

The challenges also provide children with an opportunity to practise and consolidate the five themes and objectives of Strand 1: Using and applying mathematics for Year 4 in the *Renewed Framework of Mathematics* (2006).

Creating a problem-solving classroom

It is important that children have faith in their own abilities and develop a healthy self-esteem. They need to be encouraged to have a go, even if at first their attempts are wrong. We want children to realise that having a go and making a mistake is far better than not attempting a problem at all, and that trial and improvement is a vital part of the learning process. Therefore it is important to encourage and reward the following qualities during problem-solving lessons:

- perseverance
- flexibility
- originality
- active involvement
- independence
- cooperation
- willingness to communicate and share ideas
- willingness to try and take risks
- reflection

Teacher expectations are a critical factor affecting children's achievement. We can engender a classroom ethos that makes anything possible for all children. We can offer children opportunities to reach their full potential, regardless of supposed appropriate year-level expectations.

Assessment

You can use the challenges in ***We Can Do It!*** with the whole class or with groups of children as an assessment activity. Linked to the strand that is being studied at present, ***We Can Do It!*** will not only provide you with an indication of how well the children have understood the 'pure' mathematics objectives being covered, but also their problem-solving skills.

Throughout each of the challenges there are prompting questions which focus on specific aspects of the challenge. At the end of each challenge there are also three questions that are specifically designed to help with assessing using and applying mathematics.

The list of thinking-skills statements on page 7 and the descriptions relating to the three strands of AT1 on page 8 are extremely useful in helping assess children's problem-solving skills.

Problem-solving strategies

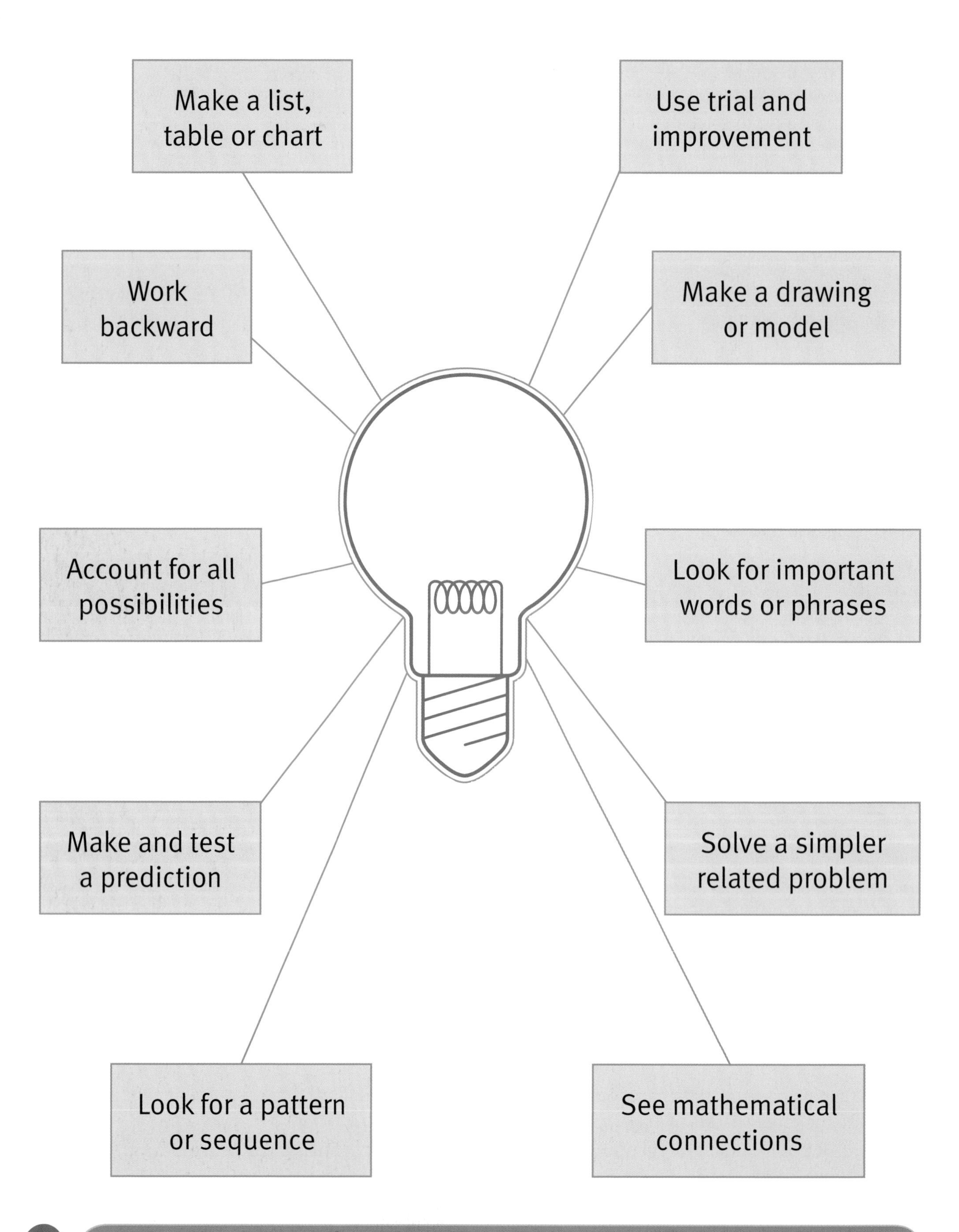

A final word

From an early age, children can learn that school mathematics is 'work' – a series of tasks they need to get through as quickly as possible, preferably without the need for thought. The challenges in ***We Can Do It!*** are deliberately demanding for children in order to promote their ability to solve problems. You will need to encourage them to rely less on your help, setting them off to work on a challenge for a short length of time. Follow this with time together to discuss the different ways in which they have set about the task; this will help children realise that they can achieve something, while you feed in ideas for continuing without taking the responsibility for the thinking away from the children.

Being challenged is enjoyable! The challenges in ***We Can Do It!*** have not been 'dressed up' to disguise the mathematics or to make them 'fun'. The aim is not to make mathematics itself enjoyable but rather find enjoyment by being prepared to have a go at something, rising to the challenge and reaching a satisfactory conclusion.

How to use this book

Question to pose the challenge

Opening question to ask the children that is designed to act as a springboard into the challenge

Summary of maths content

Brief summary of the 'pure' mathematics focus of the challenge

Introducing the challenge

Outline scenario to hook in the children's interest. It often includes opportunities to engage the children's interest further by including 'turn and talk'  instructions.

Using and applying

Description of how the challenge links to the five themes in Strand 1: Using and applying mathematics), in the *Renewed Framework for Mathematics* (2006)

Solving problems
Representing
Enquiring
Reasoning
Communicating

Maths content

Objectives from the *Framework* specifically covered in the challenge

Key vocabulary

List of words and phrases appropriate to the challenge

Resources

List of resources children need to undertake the challenge, including resource sheets (RS)

RS diagram

You find the resource sheet (RS) on the CD.

The challenge

This offers advice on how to structure the challenge and uses the following symbols for clarification:

 individual
 paired
 group
 whole class

Brick wall problem

Challenge 12

Using multiplication facts to solve a puzzle

Using and applying

Representing

Represent a puzzle using diagrams

Reasoning

Identify and use patterns, relationships and properties of numbers; investigate a statement involving numbers and test it with examples

How many different numbers are possible for the top brick?

Introducing the challenge

 Draw a brick wall with a foundation of three bricks and write the numbers 1 to 5 to one side of the wall. Children choose three of the numbers and write them in the three bricks at the bottom of the wall.

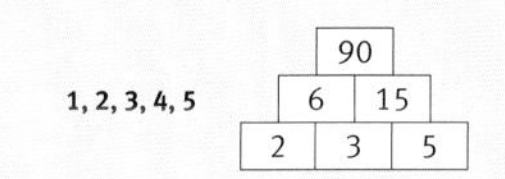

Show how this is a product wall. To work out the numbers in the bricks in the next layer up, multiply together adjacent pairs of numbers and write the product in the brick above. To find the number in the top brick, multiply together the two numbers in the middle layer. Work this out with the children, inviting answers to each calculation. Record on the bricks.

1, 2, 3, 4, 5

90
6 15
2 3 5

Repeat, starting with other numbers in the bottom three bricks.

Maths content

Knowing and using number facts

- Derive and recall multiplication facts to 10 × 10

Calculating

- Develop and use written methods to record, support and explain multiplication of two-digit numbers by a one-digit number

Key vocabulary

multiplication, product

Resources

For each pair:

- RS15 (optional)
- Pencil and paper

46 Year 4 | We Can Do It!

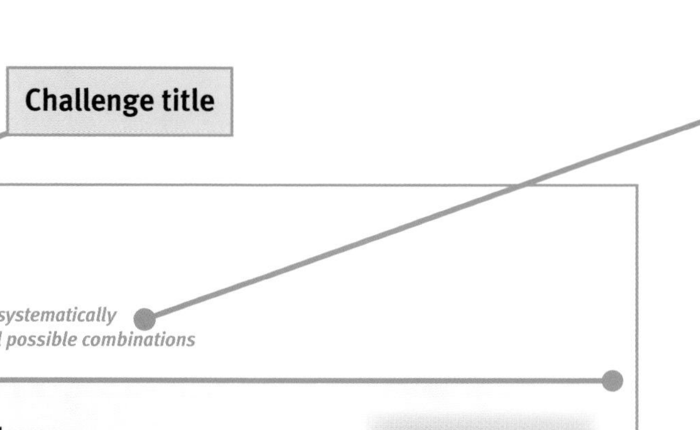

Challenge title

Summary of using and applying

Brief summary of the using and applying mathematics focus of the challenge

Working systematically to find all possible combinations

The challenge

Introduce the challenge to the children. They can use RS15 to record on. This time, they will start with a four-foundation brick wall, using the numbers 1, 1, 2 and 3 on the bottom layer. The challenge is to work out how many different numbers are possible for the top brick, using different combinations of 1, 1, 2 and 3 in the bottom row.

How do you know you have written all the different combinations?

Do you think the number on the top brick will be the same each time or different? Why?

How will you make the smallest number on top? The largest? Why does that work?

Drawing out using and applying

Invite individual pairs of children to comment on the number of different brick wall combinations there are (6, not counting mirror images) and the four different top brick products. Ask children to talk about the different strategies they used to discover the different brick wall combinations. One strategy is to consider the different centre numbers in the bottom four bricks, and then the different possible combinations for the other two numbers.

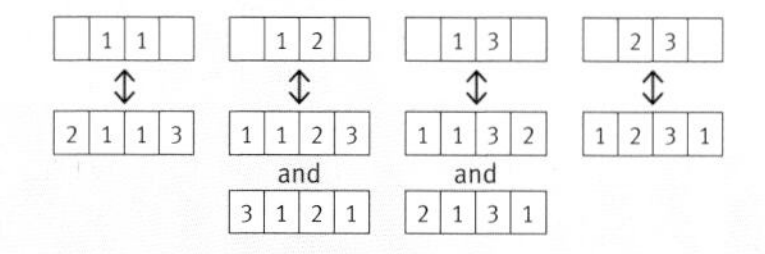

How many different combinations of brick walls are there?

How do you know you have found all the different combinations?

Did you use any strategies when you were working out how many different combinations of brick walls there were? What were they?

Did anyone use a different strategy?

What bottom row gives the smallest number on the top brick? The largest? How does that work?

What bottom rows give the same number on the top brick? Why?

Assessing using and applying

- Children can find the product in the top brick.
- Children can find different combinations of brick walls and their products.
- Children can work systematically to find all the different combinations of brick walls and their products.

Supporting the challenge

- Help children multiply a two-digit number by a one-digit number by doubling and doubling again.
- Children add (or find the difference between) pairs of adjacent numbers. How many different numbers are possible for the top brick?

Extending the challenge

- Children try different numbers on the bottom bricks.
- Children try brick walls of different sizes.

| We Can Do It! 47

Assessing using and applying

A set of three assessment questions to monitor using and applying mathematics

Supporting the challenge

Suggestions of ways to help children who are having difficulty to access the problem

Extending the challenge

Suggestions of ways to extend the challenge to cater for either the more able pupils or for the quick finishers

Drawing out using and applying

This provides advice for the plenary. It offers an opportunity to step back from the actual challenge and to think more generally about what children have learned about using and applying mathematics.

Answers

You will find the answers to some of the challenges on the CD-ROM.

Lesson suggestions

Aspects of mathematical problem solving should be covered in every maths lesson, even those that aim to teach the purest of mathematical concepts. Children need to see the application of 'pure' maths in everyday experiences.

We advise, however, that once a week you devote a lesson entirely to developing children's problem-solving skills. It is for this reason that ***We Can Do It!*** consists of 36 challenges.

The challenges in this book provide children with an opportunity to practise and consolidate the Year 4 objectives from the *Renewed Framework for Mathematics* (2006). The curriculum charts on pages 18-21 show which challenge is matched to which planning block and mathematics strand. Refer to these charts when choosing a challenge.

We Can Do It! and the daily maths lesson

The challenges in ***We Can Do It!*** are ideally suited for the daily maths lesson. You can introduce each challenge to the whole class or to groups of children. Here is a suggestion how to structure a lesson using ***We Can Do It!***.

Introducing the challenge

- Introduce the idea of the challenge either as a discussion or by giving a simplified version of the problem. You may need to highlight some of the mathematics children need to solve the problem.
- Introduce the challenge to the children by asking the question that poses the problem.
- Stimulate children's involvement through discussion.
- Use the key vocabulary throughout and explain new words where necessary.
- Make sure that the children understand the challenge.
- If you use a resource sheet, make sure children understand the text on the sheet.
- Begin to work through the challenge with the whole class, pointing out possible problem-solving strategies.

The challenge

- Arrange children into pairs or groups to work on the problem.
- Make sure appropriate resources are available to help children with the challenge.
- Monitor individuals, pairs or groups of children, offering support when and where needed.
- If appropriate, extend the challenge for some children.

Drawing out using and applying

- Plan an extended plenary.
- Discuss the challenge with the class.
- Invite individual children, pairs or groups to offer their solutions and the strategies they used.

The teacher's role in problem-solving lessons

- Give a choice where possible.

- Present the problem orally, giving maximum visual support where appropriate.
- Help children 'own the problem' by linking it to their everyday experiences.
- Encourage children to work together, sharing ideas for tackling a problem.
- Allow time and space for collaboration and consultation.
- Intervene, when asked, in such a way as to develop children's autonomy and independence.
- Work alongside children, setting an example yourself.
- Encourage the children to present their work to others.

Paired and group work

We Can Do It! recognises the importance of encouraging children to work collaboratively. All of the challenges in ***We Can Do It!*** include some element of paired or group work. By working as a group, children develop cooperation and collective responsibility. They also learn from each other, confirming their mathematical knowledge and identifying for themselves, in a non-threatening environment, any misconceptions they may hold.

The *National Curriculum* identifies three strands of the AT1: Using and applying mathematics. They are problem solving, communicating and reasoning. While it is possible for children to problem solve independently, communication, as the diagram below illustrates, is a cooperative, interactive process that involves both expressing and receiving information.

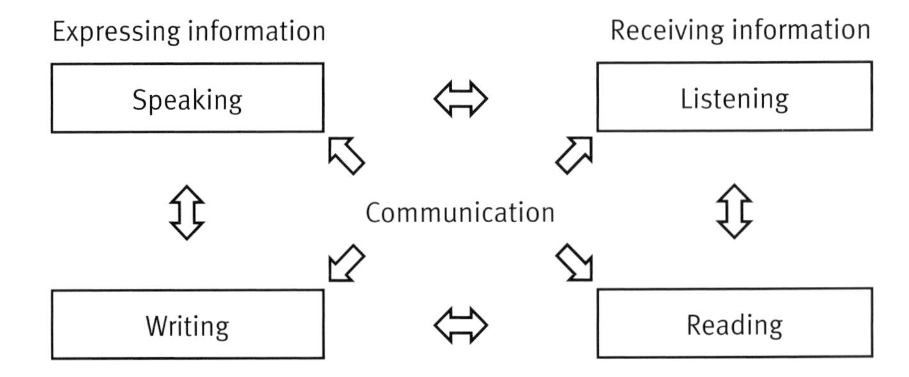

Meaningful reasoning can only occur through communication. Children cannot effectively reason with themselves: they always see themselves as being right! It is not until they begin to discuss and share ideas with others that children begin to reason and see other points of views and possibilities.

Charts linking to the *Renewed Framework for Mathematics* (2006)

Chart linking blocks and strands of the *Renewed Framework for Mathematics* (2006)

	Strand 2: Counting and understanding number	**Strand 3:** Knowing and using number facts	**Strand 4:** Calculating	**Strand 5:** Understanding shape	**Strand 6:** Measuring	**Strand 7:** Handling data
BLOCK A: Counting, partitioning and calculating	●	●	●			
BLOCK B: Securing number facts, understanding shapes		●		●		
BLOCK C: Handling data and measures					●	●
BLOCK D: Calculating, measuring and understanding shape			●	●	●	
BLOCK E: Securing number facts, calculating, identifying relationships	●	●	●			

Chart linking challenges in *We Can Do It! Year 4* to strands of the *Renewed Framework for Mathematics* (2006)

Challenge		Strand 1: Using and applying mathematics					Strand 2:	Strand 3:	Strand 4:	Strand 5:	Strand 6:	Strand 7:
Number	Title	Solving problems	Representing	Enquiring	Reasoning	Communicating	Counting and understanding number	Knowing and using number facts	Calculating	Understanding shape	Measuring	Handling data
1	Light bars				●	●	●					
2	Printed patterns				●	●	●					
3	What's in a name?				●	●	●	●				
4	Hidden number				●	●	●	●				
5	Grandma's weights	●	●				●	●				
6	It's not fair		●		●		●		●			
7	Strike out	●	●					●				
8	Turn 'em over		●		●			●				
9	Think of a number		●		●			●				
10	Number change				●	●		●				

Challenge		Strand 1: Using and applying mathematics					Strand 2:	Strand 3:	Strand 4:	Strand 5:	Strand 6:	Strand 7:
Number	Title	Solving problems	Representing	Enquiring	Reasoning	Communicating	Counting and understanding number	Knowing and using number facts	Calculating	Understanding shape	Measuring	Handling data
11	What's the rule?			●	●			●				
12	Brick wall problem		●		●			●	●			
13	Folding a number square		●			●		●				
14	Doubling trouble				●	●		●	●			
15	Dining out	●		●					●			
16	How much paper?			●		●			●			
17	Tyler's cakes	●				●			●			
18	A fraction of your money	●				●		●	●			
19	No more sixes	●			●			●	●			
20	Tables for Bingo				●	●		●	●			
21	Phone numbers	●	●						●			
22	Would you rather?	●				●		●	●			
23	Describe it				●	●				●		

Challenge		Strand 1: Using and applying mathematics					Strand 2:	Strand 3:	Strand 4:	Strand 5:	Strand 6:	Strand 7:
Number	Title	Solving problems	Representing	Enquiring	Reasoning	Communicating	Counting and understanding number	Knowing and using number facts	Calculating	Understanding shape	Measuring	Handling data
24	Designing a playground			●		●				●	●	
25	Ideal gnome home				●	●				●		
26	Two cuts	●	●							●		
27	Routes round school			●		●				●		
28	How big is an egg?	●	●								●	
29	Slow helicopter	●	●								●	
30	Making weights				●	●		●			●	
31	Cover it				●	●				●	●	
32	Bags of fun	●	●							●	●	
33	Sports day			●		●		●			●	
34	What's your name?		●	●								●
35	Books in school			●	●							●
36	How far from home?			●		●						●

The challenges

Strand 2: Counting and understanding number

Strand 3: Knowing and using number facts

Strand 4: Calculating

Strand 5: Understanding shape

Strand 6: Measuring

Strand 7: Handling data

Light bars

Challenge 1 **Reading, writing and ordering whole numbers**

Using and applying

Reasoning

Identify and use patterns, relationships and properties of numbers; investigate a statement involving numbers and test it with examples

Communicating

Report solutions to puzzles, giving explanations and reasoning orally and in writing, using diagrams and symbols

Maths content

Counting and understanding number

- Read, write and order whole numbers

Key vocabulary

digit, number, light bar, explain, justify

Resources

- Headless matchsticks
- RS1 (for each child)
- Demonstration calculator

For *Supporting the challenge*:

- Glue
- Calculator

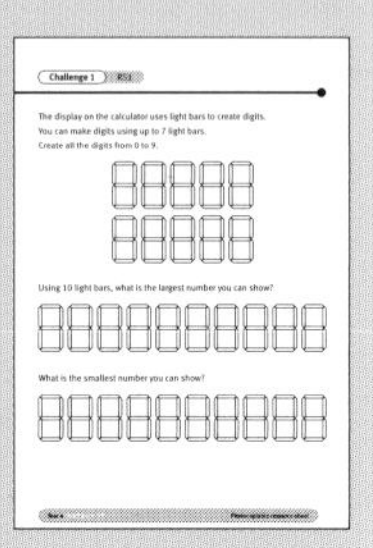

How do calculators and clocks display numbers?

Introducing the challenge

Briefly discuss with the children where they have seen numbers displayed digitally, such as cash registers, petrol pumps, digital clock displays, DVD players, and so on.

Ask the children if they have noticed how numbers are presented in digital displays.

Can anyone draw how any of the digits on the calculator screen appear?

How does each digit light up? What happens to curves such as 6 or 9?

Remind the children that we call the numbers 0 to 9 'digits', and that all the numbers in our number system are made using these 10 digits. Introduce the idea of light bars and the basic unit of seven bars from which all of the 10 digits can be displayed. Invite children to offer suggestions about how different digits are made up with light bars.

Show children the convention for displaying digits on the demonstration calculator and discuss how 4, 7 and 9, in particular, are made up.

The challenge

Provide each pair with matchsticks to use as light bars. Between them, they make all the digits from 0 to 9. They record these on RS1. They then create the largest and the smallest numbers they can in a calculator display with 10 light bars lit.

What is the difference between a number and a digit?

Which digit uses the least number of light bars? Which uses the most?

How many light bars do you need to light up to make 6? What about 9?

 Children take turns to set each other similar challenges such as:

Make a two-digit number using six light bars.

Can you make an even number using five light bars?

Make a number between 50 and 80 using eight light bars.

What is the largest number you can make using seven light bars?

Drawing out using and applying

 Discuss the results of the first challenge. Ask individual children to write up some of the light-bar digits, then invite them to show the class the same digits with the demonstration calculator.

Which digit uses six light bars? Is there another number that uses six light bars?

Discuss the second challenge. Display the numbers on the sdemonstration calculator to check.

What is the largest number you made using exactly 10 light bars?

Why isn't it possible to make a number larger than 11 111?

What is the smallest number? How do you know?

Why isn't it possible to make a number smaller than 22?

Is there any pattern to the number of light bars allowed and the largest or smallest numbers possible?

Pairs of children who worked together on problems of their own share one of their problems with the rest of the class. Allow time for the remainder of the class to work out the answer before asking the pair who presented the problem if the answer is correct. Ask other pairs of children for one of their problems and repeat several times.

Assessing using and applying

- Children can try out various numbers until they cannot find one larger/smaller.
- Children can realise that a string of 1s can be useful.
- Children can use their knowledge of place value to argue why no larger number is possible.

Supporting the challenge

- Children make all the digits with matchsticks as light bars and stick these on paper as a reference.
- Children use a calculator for reference.

Extending the challenge

- Children include the decimal point or the minus sign to show negative numbers. What is the smallest number possible they can display with a fixed number of light bars?
- Children explore how many two-digit numbers they can make with, for example, six light bars only.

Printed patterns

Challenge 2

Recognising visual patterns

Using and applying

Reasoning

Identify and use patterns, relationships and properties of numbers or shapes; investigate a statement involving numbers and test it with examples

Communicating

Report solutions to puzzles, giving explanations and reasoning orally and in writing, using diagrams

Maths content

Counting and understanding number

- Recognise and continue sequences
- Relate numbers to patterns

Key vocabulary

pattern, repeating, sequence

Resources

For each pair:

- RS2
- Coloured pencils
- Coloured counters
- RS3
- Pencil and paper

For *Extending the challenge*:

- Wallpaper and fabric

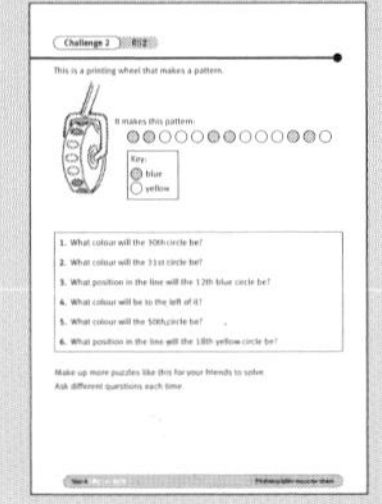

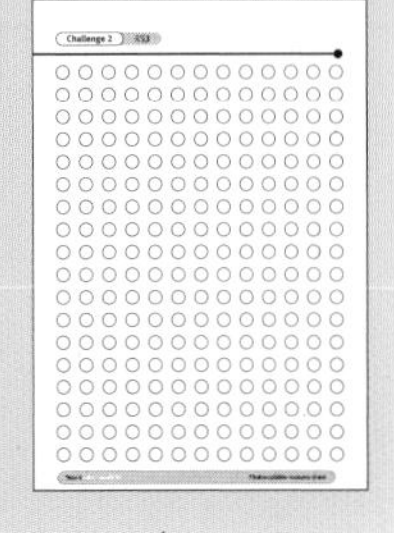

If you print a repeated pattern, can you predict colours before they are printed?

Introducing the challenge

On the board, display a simple repeating pattern such as:

 (a, b, b)

What colour is the 4th circle?

What is the position of the 4th white circle?

What colour will the next circle be? The 12th circle? The 20th circle?

How many along the line is the 10th black circle?

Describe the pattern to me. Can anyone describe it a different way?

Ask the children to explain their reasoning to the rest of the class. Also ask them to invent further questions similar to those above (but they must work out the answer first!).

Repeat the above for other simple repeating patterns.

✦✦✦⇨ (a, a, a, b)

□□❄❄□□❄❄(a, a, b, b)

The challenge

Distribute RS2 and coloured pencils to pairs of children. Briefly discuss the first part of the challenge with the children.

Would it help to make a simpler pattern?

What if you started with yellow, blue, yellow, blue ...? How would you work it out?

Why not make a guess and see what happens?

Provide children with counters to set out and count. Children can record their patterns on RS3.

For the second part of the challenge, children prepare one pattern and a set of questions and answers to present to other children (answers should be written on a separate piece of paper from the questions). You may wish to direct individual pairs of children to devise specific types of patterns to ensure an appropriate challenge: for example, simple (a, b) patterns or more complex (a, b, c) patterns.

When pairs of children have made their pattern and written their questions, they swap these with another pair whose pattern and questions have a similar degree of difficulty.

Drawing out using and applying

Discuss the answers to the first challenge on RS2.

What colour will the 50th circle be? How did you work it out?

What position in the line will the 18th yellow circle be? How do you know this?

The first pattern is in a cycle of 5, which makes calculations easier. Some children may be able to see that the 30th circle in the pattern must be yellow, because 30 is a multiple of 5. Other children may set out the counters and count them. Finding the position of a counter in the line involves multiplication or repeated addition, or a mixture of both. Encourage the children to explain their thinking.

Invite pairs of children to show their pattern and ask some of the questions they wrote for the rest of the class. Ask the children to explain why the questions are worth asking, and what strategies help to find the answers quickly.

Describe your pattern to us. Come and write it on the board.

What is one of the questions you wrote about your pattern?

Can you give us a clue about how to work out the answer to your question?

Assessing using and applying

- Children can see the patterns and relate them to number sequences to find ways of describing the sequence, such as: "This is a 2, 3, 2, 3 pattern."
- Children can see the pattern as a chunk of, for example, 5 that repeats itself and use this to predict the numbers.
- Children can make explicit statements such as: "This pattern goes in 5s, and I need to get to the 30th circle. I know that 5 goes into 30, so I know the pattern must end at 30 and must be yellow."

Supporting the challenge

- Help children solve the problem on a one-to-one basis if necessary.
- Give children a simpler starting problem such as those used when introducing the challenge.

Extending the challenge

- Children use three colours or patterns such as 3, 2, 5, 2, 3, 2, 5, 2.
- Children look at patterns on wallpaper and fabric and explore the repeats. They work out a code for the patterns.

What's in a name?

Exploring and explaining number patterns and recognising multiples

Challenge 3

Using and applying

Reasoning

Identify and use patterns, relationships and properties of numbers; investigate a statement involving numbers and test it with examples

Communicating

Report solutions to problems, giving explanations and reasoning orally and in writing, using diagrams and symbols

Maths content

Counting and understanding number

- Recognise and continue number sequences formed by counting on or back in steps of constant size

Knowing and using number facts

- Derive and recall multiplication facts to 10×10, the corresponding division facts and multiples of numbers to 10 (to the tenth multiple)

Key vocabulary

count, pattern, relationship, multiplication, division, multiple

Resources

- NNS ITP: Number grid
- RS4 (for each child)
- Coloured pencils (for each group)

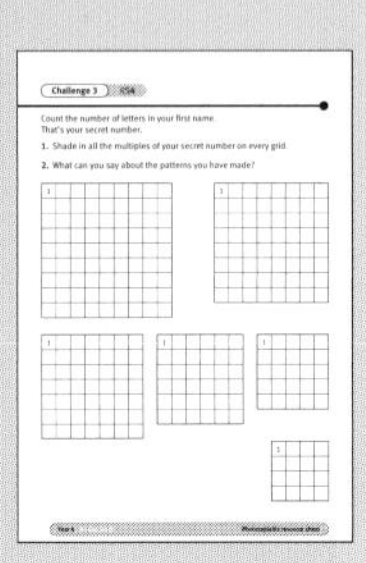

By looking at number patterns, can you guess the name of the person who made them?

Introducing the challenge

Ask the children to imagine a grid of squares that is six squares wide. Ask them to imagine the number 1 in the top left-hand corner of the grid, the number 2 to the right of it, and so on, to 6.

Where will the number 7 be?

What about the number 12? 21? 30?

How did you work it out?

Using the NNS ITP: Number grid, display a 6-column grid starting from 1. Mask all the numbers on the grid with the exception of the first row (numbers 1 to 6). Ask questions similar to the above, revealing each number after the children have provided an answer.

If appropriate, repeat the above for other grid sizes.

The challenge

Give out RS4. Each child uses the number of letters in their first name as their secret number. (Children can shorten their name if it is too long to manage.) They shade in the multiples of their secret number on each grid. For example, if a child's name is David and they therefore have 5 as their secret number, they will shade in the squares that would contain 5, 10, 15, 20 ... wherever these occur.

1					

1			

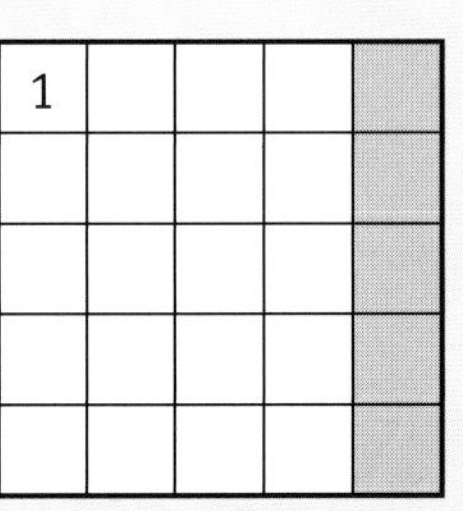

Groups shuffle their completed resource sheets and swap with another group. Can they work out from the patterns on the grids which possible name (or names) were used for each pattern?

Encourage children to look for patterns. Can they spot the connection between different grid sizes and multiples for any particular pattern? For example, they might see that there is a diagonal strip when the grid width is one less than the multiple (the number of letters in a name).

Why are all these grids similar? What is the same about them?

Is there anything slightly different about them?

What patterns do you notice?

Can the children identify any relationships between a grid size, the multiple shaded and the pattern created? For example, multiples of 3 will form vertical stripes on both the 6-grid and the 9-grid. Can they predict what will be the next-size grid to create vertical stripes for multiples of 3?

What do you notice about the different patterns on the same grids? On different grids?

Which other grids have a vertical or diagonal pattern? What is the same about the numbers chosen to make the pattern?

Can you predict what will be the next-size grid to make the same pattern?

If you start with 4, what pattern would you make on a 10-grid? What about a 12-grid? What about if you started with 7?

Drawing out using and applying

Groups choose one particular pattern and write down everything they have found out about it. Ask each group to read out their statements and encourage the other children in the class to offer further suggestions to improve the quality of the explanation and add more information.

Who else wrote a statement about the multiples of 3? What else can you tell us about them? Can anyone else tell us something different?

Who can tell us something about the multiples of 6?

Assessing using and applying

- Children can describe the various patterns accurately and use appropriate mathematical language.
- Children can identify and describe the connection between a particular multiple, the grid size and the patterns created.
- Children can talk and write about the patterns clearly and unambiguously.

Supporting the challenge

- Give children easier multiples to work with, such as 2, 3, 4 or 5. They can choose shorter names for themselves.
- Children write in the number on the grid before shading in the multiples of their name.

Extending the challenge

- What patterns can you make if you shade two different sets of multiples on the same grid?
- What happens if the grid numbering does not start at 1?

Hidden number

Exploring and explaining number patterns and recognising multiples

Challenge 4

Using and applying

Reasoning

Identify and use patterns, relationships and properties of numbers; investigate a statement involving numbers and test it with examples

Communicating

Report solutions to problems, giving explanations and reasoning orally and in writing, using diagrams and symbols

Given the pattern on the grid, can you figure out the chosen number?

Introducing the challenge

Children need to have done Challenge 3 before doing this challenge.

Ask the children to imagine a grid of squares that is four squares wide. Ask them to imagine the number 1 in the top left-hand corner of the grid, the number 2 to the right of it, and so on, to 4.

Where will the number 5 be?

What about the number 9? 10? 12?

How did you work it out?

Ask the children to imagine that all the multiples of 2 are shaded in on the grid.

What pattern would they form?

Using the NNS ITP: Number grid, display a 4-column grid starting from 1. Highlight all the multiples of 2 to reveal the vertical strip pattern.

1	2	3	4
5	6	7	8
9	10	11	12
13	14	15	16
17	18	19	20
21	22	23	24

Repeat the above, asking the children to imagine a grid of squares that is 6 squares wide. Ask the children to imagine that all the multiples of 3 are shaded in on the grid. However, this time, do not show the pattern on the interactive whiteboard.

The challenge

Give out RS5. Children check their predictions for the patterns of 3 on the 6-column grid.

Was your prediction correct?

What pattern has it made?

Maths content

Counting and understanding number

- Recognise and continue number sequences formed by counting on or back in steps of constant size

Knowing and using number facts

- Derive and recall multiplication facts to 10×10, the corresponding division facts and multiples of numbers to 10 (to the tenth multiple)

Key vocabulary

count, pattern, relationship, multiplication, division, multiple, width

Resources

- NNS ITP: Number grid

For each child:

- RS5, blank square grid

For *Supporting the challenge*:

- RS6

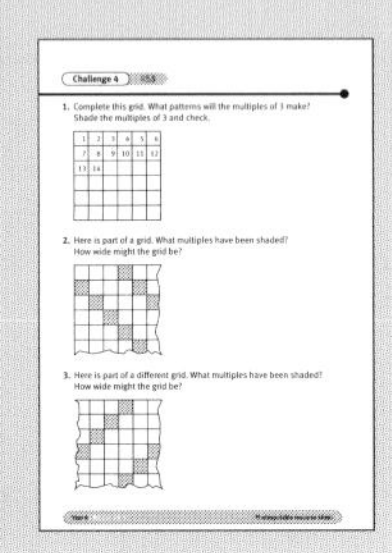

The unnumbered grids are the main challenge on the resource sheet. Explain to the children that each of these grids has been torn and the numbers from the grid removed. The squares that contain multiples of a particular number are still shaded. Tell the children that, for each grid, they have to work out which multiples have been shaded, and how wide each grid is.

Which multiples have been shaded? How do you know?

How wide do you think the grid is?

Is this the only size grid it could be?

What other width might the grid be?

Four is the only possible value for the multiple on the first blank grid, but there are several possible answers for the width. For example, a width of 7 or 11 will result in the same pattern. Five is the only possible value for the multiple on the second blank grid; again, there are several possible answers for the width.

Encourage the children to work together and discuss possible solutions before drawing up a grid to check them.

Once the children have identified the multiples and grid width for both grids, set them off to create grid problems for each other to solve.

Drawing out using and applying

Discuss and compare the strategies the children used to solve the different grid problems they devised.

How do you know which multiples were shaded in?

How wide do you think the grid is? Is this the only size grid it could be? What other size might the grid be?

What other grid widths could they be? What do you notice about all these grids?

Invite children who made their own grid problems to show these to the class. If possible, use the ITP to display these and ask the rest of the class to identify the multiple and the possible grid widths.

Assessing using and applying

- Children can spot the grid width patterns of 7, 11, 15 ... and 6, 12, 18 ... for the problems provided and describe similar patterns for their own problems.
- Children can devise a strategy that will enable them to solve any grid problem efficiently.
- Children can begin to solve grid pattern problems where two multiples have been shaded.

Supporting the challenge

- Provide the children with some of the completed grids from Challenge 3.
- Provide the children with a copy of RS6 to try out different possibilities.

Extending the challenge

- What happens if the part of the grid you see is not the top of the grid but taken from the middle of the grid?
- What happens if you shade in two different sets of multiples on the same grid?

Grandma's weights

Challenge 5 | **Solving a problem and identifying patterns**

Using and applying

Solving problems

Solve problems involving measures

Representing

Represent a problem using number sentences, statements or diagrams; use these to solve the problem; present and interpret the solution in the context of the problem

Could you weigh everything from 1 oz to 32 oz, using just five weights?

Introducing the challenge

Discuss with the children the different units of measure they use to weigh objects (grams and kilograms) and explain that this measuring system is called the 'metric system'.

Introduce the imperial system of measures. Explain how imperial weights were used in the UK until relatively recently. Point out that children may now only use grams and kilograms to weigh objects, but that many people still think in pounds and ounces.

What unit of measure do you use to measure weight?

What do you know about the imperial system of measures?

Do we still measure anything using the imperial system?

The challenge

If appropriate, provide each pair with a copy of RS7. Show the children the 1 oz, 2 oz, 4 oz, 8 oz and 16 oz weights. Present the challenge to the children and explain that they need to find out if Grandma is correct. They must work out solutions to weighing out every amount from 1 oz to 32 oz, using only combinations of some or all of the five weights: 1 oz, 2 oz, 4 oz, 8 oz and 16 oz.

As they work through the challenge, encourage the children to think of more than one solution for each weight. For example, they can make 6 oz using the 2 oz and 4 oz weights, but also by using the 8 oz weight and the 2 oz weight. Discuss how children might do this practically: they weigh out 8 oz of flour with the 8 oz weight, then replace the 8 oz weight with the 2 oz weight. They carefully spoon out the flour into a bowl until 2 oz is left. The flour in the bowl weighs 6 oz.

How did you work out how to weigh 6 oz? 17 oz? 23 oz?

Is there another way?

How are you going to record your findings?

Maths content

Counting and understanding number

- Recognise number sequences

Knowing and using number facts

- Use knowledge of addition and subtraction facts and place value to derive sums and differences

Key vocabulary

number, weight, sequence, pattern, possibilities, metric, imperial, ounce (oz)

Resources

- 1 oz, 2 oz, 4 oz, 8 oz and 16 oz weights (optional)
- RS7 (optional, for each pair)

For *Supporting the challenge*:

- Interlocking cubes
- Squared paper

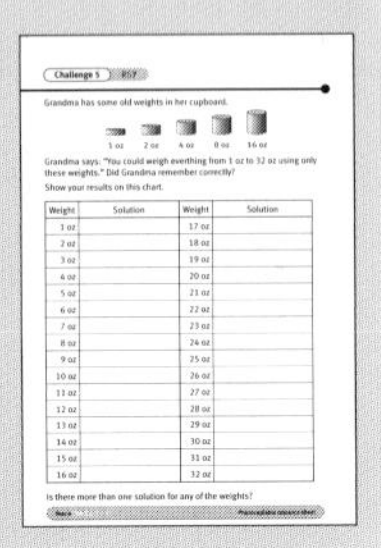

Challenge 5 RS7

Grandma has some old weights in her cupboard.

1 oz 2 oz 4 oz 8 oz 16 oz

Grandma says: "You could weigh everthing from 1 oz to 32 oz using only these weights." Did Grandma remember correctly?

Show your results on this chart.

Weight	Solution	Weight	Solution
1 oz		17 oz	
2 oz		18 oz	
3 oz		19 oz	
4 oz		20 oz	
5 oz		21 oz	
6 oz		22 oz	
7 oz		23 oz	
8 oz		24 oz	
9 oz		25 oz	
10 oz		26 oz	
11 oz		27 oz	
12 oz		28 oz	
13 oz		29 oz	
14 oz		30 oz	
15 oz		31 oz	
16 oz		32 oz	

Is there more than one solution for any of the weights?

Photocopiable resource sheet

Drawing out using and applying

Bring the class back together and discuss the various solutions children found.

How did you work out how to weight 5 oz? 15 oz? 27 oz?

What patterns did you notice?

Who used trial and improvement?

Did you see a system for finding all the solutions?

Discuss the various ways children recorded their solutions.

How did you write down your solutions?

Did anyone write them down using a different method?

How might you have recorded your solutions differently? Might this have been better? How?

Did making a list help you see any patterns?

Finally, ask the children whether or not Grandma remembered correctly: that you could weigh everything from 1 oz to 32 oz using just the 1 oz, 2 oz, 4 oz, 8 oz and 16 oz weights. (It is possible to make weights from 1 oz to 31 oz, but not 32 oz.)

Did Grandma remember correctly?

What do you think was the next standard ounce weight? (32 oz)

Assessing using and applying

- Children can work in an unsystematic fashion, finding some solutions and then filling the gaps, using trial and improvement.
- Children can work in a systematic way to find all the solutions.
- Children can work in a systematic way and explain how the pattern of weights builds up.

Supporting the challenge

- Children use interlocking cubes to model the problem. They work with five lengths of cubes only: 1, 2, 4, 8, 16.
- Children record by outlining or shading squared paper to represent the cubes.

Extending the challenge

- Children find the metric equivalents of all the weights from 1 oz to 32 oz.
- Children make imperial weights and make a poster showing the rough equivalents between pounds and kilograms, and ounces and grams.

It's not fair

Challenge 6 **Finding fractions of numbers and quantities**

Using and applying

Representing

Represent a problem using diagrams

Reasoning

Identify and use patterns, relationships and properties of numbers

How can you share out the food?

Introducing the challenge

Organise children into groups of three and ask them to work out how they would share each of the following items between the three of them: 12 sweets, a pizza, 4 apples, 8 cakes, £2.91 and 1 litre of orange juice.

Discuss some of the children's solutions.

How many apples did you get each?

What did you do about the 8 cakes?

The challenge

Give the groups RS8 and explain the challenge. Children record their solutions, using calculations or drawings where appropriate.

Invite children to explain their solutions as they are working and to show their methods of recording. Look at examples that are difficult to share accurately in real life.

How would you share the litre of ice cream?

Would it be possible to measure out these amounts of ice cream as accurately as the rule says?

Can you give fair shares in real life without being absolutely accurate?

Show the children how to draw out a box made up of 8 equal parts (or they could fold a sheet of A4 paper into 8 equal parts). Colour $\frac{4}{8}$, $\frac{2}{8}$ and two separate $\frac{1}{8}$s and relate the sizes of the boxes to the shares needed by the Shakespeare family. Show the children how they can use this to help solve the problem.

Their next challenge is to work with a partner and suggest other things that the family might share.

Maths content

Counting and understanding number

- Use diagrams to identify equivalent fractions

Calculating

- Find fractions of numbers and quantities

Key vocabulary

fraction, whole, half, quarter, eighth, third, sixth, tenth, equivalent, numerator, denominator, division, share

Resources

For each group:

- RS8
- A4 paper

For *Supporting the challenge*:

- Counters, paper, coins, bottles
- Jugs, water
- Squared paper

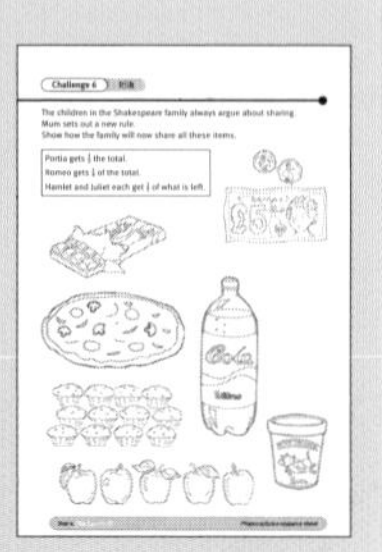

Drawing out using and applying

Display the finished work and ask children to explain how certain quantities can be shared.

How would you share out 1 kg of peanuts? Would you count the nuts or weigh them?

How can you share out liquids accurately?

How can you cut a pizza into fair shares?

Assessing using and applying

- Children can work out simple fractions of numbers and quantities and represent these in diagrams.
- Children can work out simple fractions of numbers and quantities, recognise equivalents such as $\frac{2}{4}$ and $\frac{1}{2}$, and $\frac{2}{8}$ and $\frac{1}{4}$, and represent these, using diagrams, numbers and fraction notation.
- Children can solve problems with fractions, recognising the whole the fraction is part of, and the equivalences of the fractions (such as $\frac{3}{12}$ and $\frac{1}{4}$), and represent their solutions, using diagrams, numbers and fraction notation.

Supporting the challenge

- Children use apparatus such as counters, paper, coins and bottles and jugs of water to model the quantities.
- Offer suggestions about how children record their findings, including folding paper, using squared paper and drawing objects.

Extending the challenge

- Children work out how three children could divide the amounts on the resource sheet if they share things $\frac{1}{2}$, $\frac{1}{3}$, $\frac{1}{6}$.
- Children work out a rule for sharing things fairly in the class.

Strike out

Challenge 7

Calculating mentally, using all four operations

Using and applying

Solving problems

Solve problems involving numbers; choose and carry out appropriate calculations

Representing

Represent a problem using number sentences

Maths content

Knowing and using number facts

- Use knowledge of addition and subtraction facts and place value to derive sums and differences
- Derive and recall multiplication facts to 10×10 and the corresponding division facts

Key vocabulary

addition, subtraction, multiplication, division, score, calculation, number sentence

Resources

- Individual whiteboard and pen (for each child)

For each pair:

- RS9
- RS Extra (digit cards: 1, 1, 1, 4, 4, 4, 9, 9, 9)
- Bag or box

For *Supporting the challenge*:

- RS Extra (number cards: 1, 1, 1, 2, 2, 2, 5, 5, 10)
- Number lines

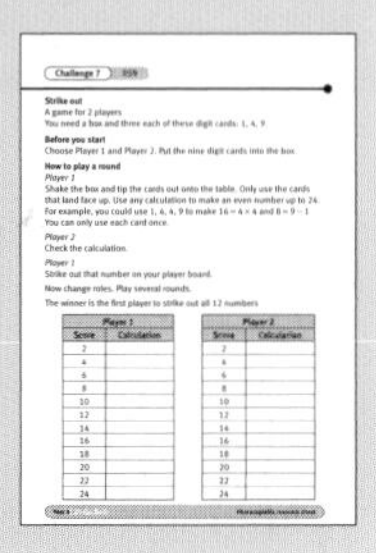

Can you make all the even numbers from 2 to 24, using only ones, fours and nines?

Introducing the challenge

Introduce the game 'Strike out' to the class. Show children the nine digit cards (ones, fours and nines, see RS Extra) and place them in a bag or box. Shake them up and tip them out onto a table. Invite a child to say the numbers that are face up and write these on the board: for example, 1, 4, 4, 9.

Children use some or all of the numbers that have landed face up and any of the four operations to make an even number. For example:

$36 = 4 \times 9$

$18 = 9 + 4 + 4 + 1$

$16 = 4 \times 4$

$12 = 9 + 4 - 1$

$4 = 4 \times 1$

$2 = (9 - 1) \div 4$

Take several suggestions from the class and write these on the board.

The challenge

Children play the game in pairs, using RS9 and the cards from RS Extra. Encourage them to help each other. Remind them that they do not have to use all the cards that have landed face up. Talk about using all the four operations and brackets, and not just addition and subtraction.

How can you use these cards to make an answer of 12?

Can you use these same cards to make another even number less than 24?

Can you think of another calculation that uses these numbers and also has the answer of 18?

What could you change for your answer to be an even number less than 24?

Drawing out using and applying

Invite individual children to say some of the calculations they created. Write some of the more complex calculations on the board and discuss these with the class, in particular the ones that use multiplication and division.

Which answers were easy to strike out?

Which were not so easy? Why?

Which combinations of number cards were good to have landing face up?

Which were not so good? Why?

What was the longest calculation you thought of?

Who can tell me one of their calculations that used division? More than two numbers? Brackets?

Assessing using and applying

- Children can combine the numbers using only addition and subtraction.
- Children can combine numbers using all four operations.
- Children can select particular numbers from the list at which they target their solutions.

Supporting the challenge

- Make sure children have number lines and other supporting equipment available.
- Use RS Extra with the following number cards: 1, 1, 1, 2, 2, 2, 5, 5, 10.

Extending the challenge

- Children decide on their own set of number cards.
- From one toss of the cards on the table, how many different totals can everyone make?

Turn ’em over

Challenge 8

Calculating with the four operations mentally

Using and applying

Representing

Represent a puzzle using number sentences; use these to solve the problem

Reasoning

Identify and use patterns, relationships and properties of numbers; investigate a statement involving numbers and test it with examples

Maths content

Knowing and using number facts

- Use knowledge of addition and subtraction facts and place value to derive sums and differences
- Derive and recall multiplication facts to 10 × 10 and the corresponding division facts

Key vocabulary

addition, subtraction, multiplication, division, combination

Resources

- 21 blank cards (for each child)

For each pair:

- Individual whiteboard and pen, two 1–6 dice, RS10

For *Supporting the challenge*:

- Two 1, 2, 3, 1, 2, 3 numeral dice

For *Extending the challenge*:

- Two 6–12 dice or two 0–9 dice

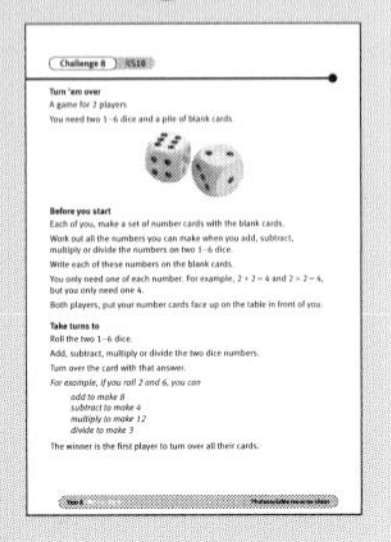

Challenge 8 RS10

Turn ’em over

A game for 2 players

You need two 1–6 dice and a pile of blank cards.

Before you start

Each of you, make a set of number cards with the blank cards.

Work out all the numbers you can make when you add, subtract, multiply or divide the numbers on two 1–6 dice.

Write each of these numbers on the blank cards.

You only need one of each number. For example, 2 + 2 = 4 and 2 × 2 = 4, but you only need one 4.

Both players, put your number cards face up on the table in front of you.

Take turns to

Roll the two 1–6 dice.

Add, subtract, multiply or divide the two dice numbers.

Turn over the card with that answer.

For example, if you roll 2 and 6, you can

add to make 8
subtract to make 4
multiply to make 12
divide to make 3

The winner is the first player to turn over all their cards.

If you roll two dice, what are all the numbers you can make if you add, subtract, multiply or divide the two numbers?

Introducing the challenge

Provide each pair of children with an individual whiteboard and pen and two 1–6 dice. Discuss a simpler version of the game on RS10 with the children. However, at this stage of the lesson, the children are not going to play the game.

- Each child has a set of cards with numbers, and each pair of children has two 1–6 numeral dice. Each child places their set of cards face up on the table in front of them.
- Children take turns to roll the two dice and add or subtract the two numbers rolled to find their score. They turn over the number card with that answer on it.
- The first person to turn over all their number cards is the winner.

Children work in pairs to decide what numbers need to be on their set of cards for the game to work. They write these numbers on their whiteboard. (They will need the numbers 0 to 12.)

How did you work out what these numbers were?

How do you know that you have got all the numbers you need to play the game?

The challenge

Give each pair a copy of RS10 and each child the blank cards. Read through and discuss the game with the children. Emphasize that, with this version of the game, they need to work out what answers are possible when you multiply and divide the two dice numbers as well as add and subtract. (Each child needs the following 21 number cards: 0, 1, 2, 3, 4, 5, 6, 7, 8, 9, 10, 11, 12, 15, 16, 18, 20, 24, 25, 30, 36.)

Encourage the children to be systematic in their work and to record their strategies for finding the numbers for the cards as they go. When they have decided on and made their set of number cards, they play the game. Check that each child has identified and made the correct 21 number cards beforehand.

You already know you need the numbers 0 to 12 for addition and subtraction. What other numbers will you need for multiplication and division?

How many numbers are there this time?

How did you work out what the answers to the multiplication and division would be?

Have you got all the numbers you need to play the game? How can you be sure?

Drawing out using and applying

Ask individual children to share with the rest of the class how they worked out the numbers for the cards, and what systems they used to make sure they found them all.

How did you keep track of what you were doing?

Did you organise the numbers in any way?

Did anyone use a different system? Tell or show us.

Finally, discuss the results of the game with the children.

Were there some numbers that were more difficult to score for than others? Why was that?

Assessing using and applying

- Children can work through each operation systematically, pairing the numbers on the dice in turn.
- Children can work back from a number such as 13 or 18 and decide if any operations would make up that number.
- Children can work methodically and record their findings in an organised way.

Supporting the challenge

- Children play the addition and subtraction version of the game.
- Children roll two 1, 2, 3, 1, 2, 3 numeral dice.

Extending the challenge

- Children roll two 6–12 dice or two 0–9 dice.
- Children turn over as many different cards as there are possible scores: for example, with the numbers 2 and 6, they turn over 8, 4, 12 and 3.

Think of a number

Investigating the inverse relationships between the four operations

Challenge 9

Using and applying

Representing

Represent a puzzle using number sentences, statements or diagrams; use these to solve the problem; present and interpret the solution in the context of the problem

Reasoning

Identify and use patterns, relationships and properties of numbers; investigate a statement involving numbers and test it with examples

Maths content

Knowing and using number facts

- Use knowledge of addition and subtraction, and multiplication and division facts
- Use knowledge of number operations and inverses to estimate and check calculations

Key vocabulary

addition, subtraction, multiplication, division, doubling, twice, halving, inverse

Resources

- RS11 (for each pair)

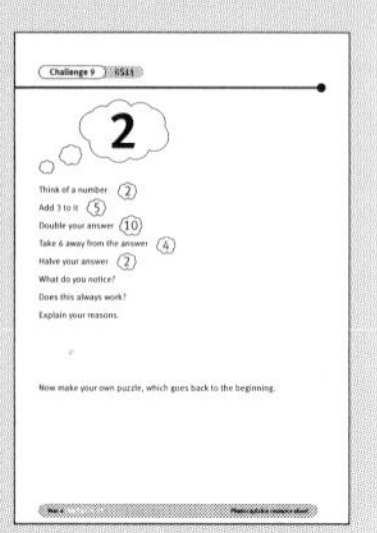

Can you create a magic 'think of a number' that goes back to the starting number?

Introducing the challenge

Write the following on the board:

Think of a number.
Add 5 to it.
Subtract 3 from your answer.
Add 2 to your answer.s
Subtract 4 from your answer.

Tell the children to think of a number and go through the steps on the board.

What do you notice?

Did everyone get back to the number that they started with?

Repeat the above for one or both of the following 'think of a number' puzzles.

Think of a number.
Double it.
Add 4 to your answer.
Divide your answer by 2.
Subtract 2 from your answer.

Think of a number.
Multiply it by 10.
Halve your answer.
Divide your answer by 5.

The challenge

Provide each pair of children with a copy of RS11. Tell the children to work through the 'think of a number' puzzle on the sheet. Encourage them to think about why they always get back to the original number.

Why does this puzzle go back to the number you started with?

Does this always work? What about much larger numbers?

It is helpful to draw a cloud for the number they are thinking of. So they start off with ☁. It then becomes ☁ + 3. Doubling the answer gives 2 × ☁ + 6 (doubling ☁ gives two times ☁ and doubling the + 3 turns into + 6). Taking away 6 leaves 2 × ☁, and halving this brings you back to the start: that is, ☁.

Think of a number.	☁
Add 3 to it.	☁ + 3
Double your answer.	2 × (☁ + 3)
Take away 6 from your answer.	(2 × ☁ + 6) − 6
Halve your answer.	(2 × ☁) ÷ 2 = ☁

Once the children can explain the reasoning behind the puzzle, they make up their own 'think of a number' puzzle, ending up with the starting number.

Drawing out using and applying

Look at the 'think of a number' puzzle on RS11. Ask children to explain why this always works.

Louisa, can you explain to us why this always works?

Who can explain it differently?

Did anyone make this work using a different method?

Show and describe how to use symbols or shapes to represent the numbers. Discuss how inverse relationships work in each of these puzzles.

Go back to the puzzle you wrote on the board. Invite children to suggest symbols or shapes.

Finally, ask some of the children to present their puzzles to the rest of the class.

How did you work out this puzzle? What did you do first?

Assessing using and applying

- Children can test that the method given works for several numbers.
- Children can make up similar simple two-step puzzles.
- Children can explain why these simple puzzles work.

Supporting the challenge

- Provide the children with simple two-step or three-step puzzles to solve and ask them to write similar puzzles. For example:

  Think of a number.
  Double it.
  Divide your answer by 2.

  Think of a number.
  Add 5 to it.
  Subtract 6 from your answer.
  Add 1 to your answer.
- Children create puzzles involving addition and subtraction, and doubling and halving.

Extending the challenge

- Children create a function machine that, for example, adds 5. Given the numbers that come out of the machine, can the children work out what numbers went in?
- Can the children find two different methods for halving large numbers?

Number change

Challenge 10

Investigating patterns in inverse operations

Using and applying

Reasoning

Identify and use patterns, relationships and properties of numbers; investigate a statement involving numbers and test it with examples

Communicating

Report solutions to puzzles, giving explanations and reasoning orally and in writing, using diagrams and symbols

Maths content

Knowing and using number facts

- Use knowledge of addition and subtraction, and multiplication and division facts
- Use knowledge of number operations and inverses to estimate and check calculations

Key vocabulary

addition, subtraction, multiplication, division, doubling, twice, halving, inverse

Resources

For each child:

- RS12
- RS13

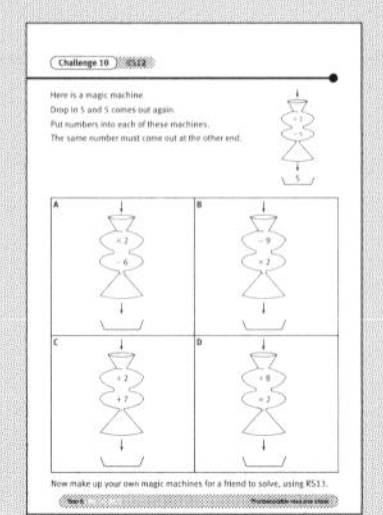

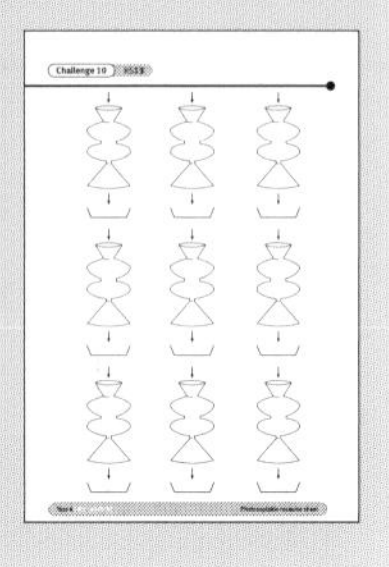

What numbers come out of the magic machines unchanged?

Introducing the challenge

Introduce a magic calculating machine that has two parts to it that change numbers in some way. Draw the machine and put in 4, then take it through $\times 2$ and -4, so that 4 comes out at the other end. Invite children to give you different numbers. For each one, create two operations that will give the starting number at the end. Draw these as you go and keep them on the board. Invite children to notice what has happened to the numbers.

Children invent some of these themselves. Write these up.

Go through one or two examples where you know what happens in each part of the machine, but you don't know which number goes in and comes out the same.

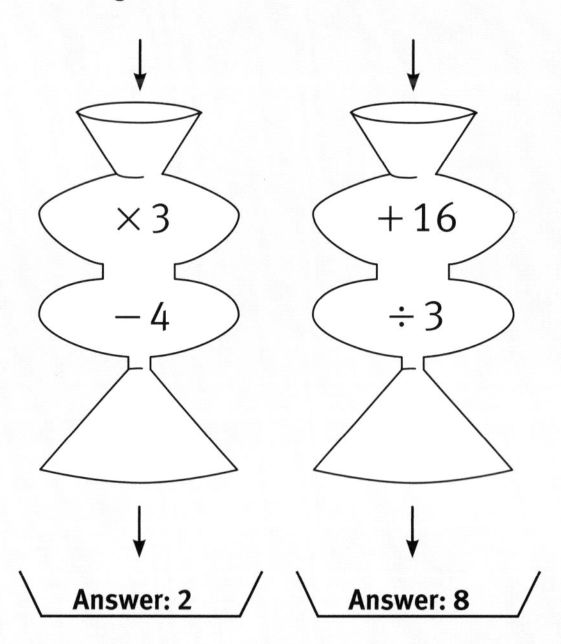

Ask the children to offer possible numbers and discuss strategies that help identify the correct number to be put in.

How did you work out what number went into the machine?

The challenge

The first part of the challenge is to solve the specific problems on RS12. The children keep a record of all the numbers they tried and what results they had.

The second part is to create their own problems for a friend to solve, using RS13. Encourage the children to make these as challenging as they can, using a mixture of operations. Remind them to work out their own answers and record them to refer to later.

Drawing out using and applying

Children share their strategies for working out the problems with the whole class.

How do you know it is the number 6 that went into the machine?

Does it make a difference whether you multiply by 2 first or last?

Does it make a difference whether you divide by 2 first or last? What happens?

What difference does it make to the numbers you try?

Did you notice any patterns in the numbers you tried?

Could you chart your results in a table?

Encourage children to explain their methods for finding the unchanged number, even if they 'just know' what it must be. This will lead to a discussion about reversing operations to 'undo' them.

Ask individual children to present one of their own problems to the rest of the class.

How did you invent your problem?

Did you start with the number first or the operations? Why?

Assessing using and applying

- Children can find the unchanged number and explain how they got it, such as: "You just double the number you take away."
- Children can invent problems that make sense for others.
- Children can make general statements about how to work out the problems, such as: "You do the opposite of what it tells you to do to find the answer, so if it says divide, you multiply."

Supporting the challenge

- When using trial and improvement, encourage the children to be systematic in the numbers they choose and to record these.
- Children use addition and subtraction of numbers to 10 and multiplication and division facts for the 2, 5 and 10 times tables.

Extending the challenge

- Children develop the idea of function machines in all kinds of different ways. They investigate three-step problems

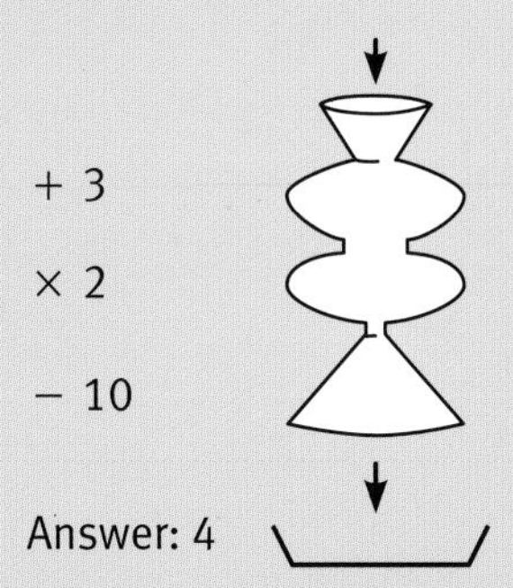

- Children look at one set of functions and at the patterns they get with different numbers.

What's the rule?

Challenge 11

Recognising common multiples

Using and applying

Enquiring

Suggest a line of enquiry and the strategy needed to follow it; collect, organise and interpret selected information to find answers

Reasoning

Identify and use patterns, relationships and properties of numbers; investigate a statement involving numbers and test it with examples

Maths content

Knowing and using number facts

- Derive and recall multiplication facts to 10×10 and the corresponding division facts
- Derive and recall multiples of numbers to 10 (to the tenth multiple)

Key vocabulary

multiple, common multiple, times table, multiplication, division

Resources

- RS14 (one for each pair and one for each group)
- RS Extra multiplication grid (for each child)

For *Supporting the challenge*:

- Coloured pencils

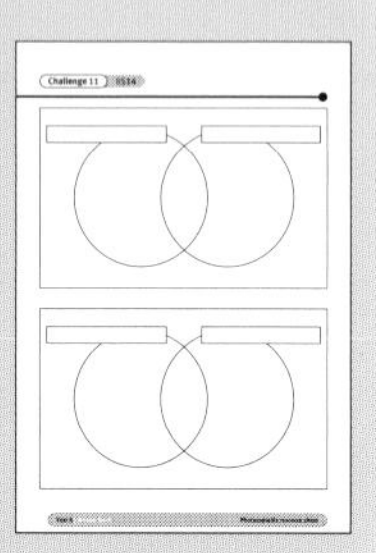

Can you work out what the rule is for sorting out the numbers?

Introducing the challenge

Play the 'Yes/No' game. Draw a circle on the board and label it 'Yes'. Draw a box and write 'No' inside it. Invite children to suggest numbers to 50. Without saying anything, place the multiples of 5 in the circle and the rest of the numbers outside.

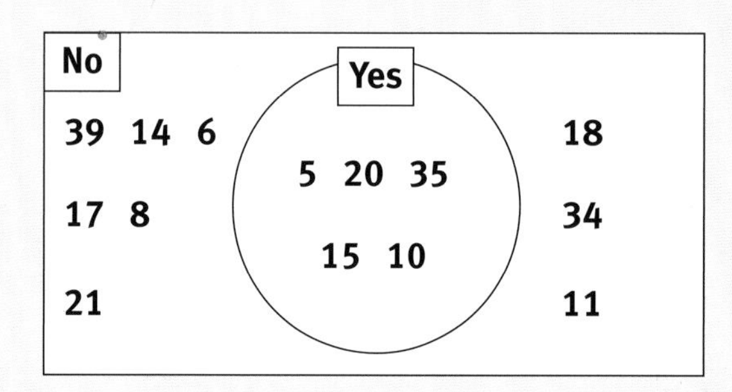

Children guess your rule for 'Yes'.

Change the diagram to two overlapping circles (an intersecting Venn diagram). Again, children suggest numbers and then guess your rule according to where you put the numbers. The aim is to get as many numbers in the circles as possible. Ask children who think they have guessed not to say what the rule is, but to try and suggest numbers that will check the rule. This time, choose multiples of 2 and 3 and multiples of 6 in the intersection.

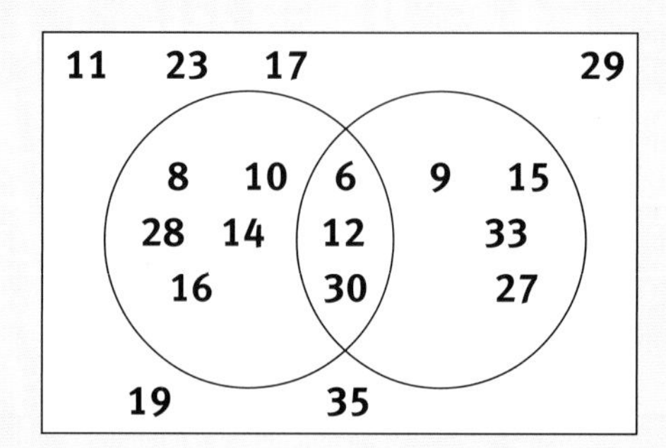

When the class thinks they have guessed the rules, ask them to explain their thinking.

Label the two circles 'Multiples of 2' and 'Multiples of 3'. Ask children to describe the numbers in the intersection. They may say that they are both multiples of 2 and of 3, and that they are multiples of 6. Introduce the term 'common multiples' and explain that the numbers in the overlap are common multiples of 2 and 3.

The challenge

The challenge is to play the 'Yes/No' game with another pair of multiples. Pairs prepare their own Venn diagram by choosing two sets of multiples, one for each circle, using RS14. They use the RS Extra multiplication grid to check the multiples. They fill in the first diagram, including labelling the two circles.

Pairs challenge each other with the 'Yes/No' game, using a blank copy of RS14. The pair offering the challenge uses their prepared diagram to check the numbers offered by the guessing pair. When the guessing pair has worked out the rule, the pairs swap roles.

The aim is to guess the rule with the fewest numbers being placed in the outer box. This means focusing on the patterns of the multiples by making reasoned guesses, rather than making random guesses.

Drawing out using and applying

Make a list of the pairs of numbers children have chosen for the 'Yes/No' game. Encourage the class to help produce the common multiples for each of these pairs of numbers.

What are the common multiples of 3 and 5?

Are there any more? How do you know?

We have 30 here as a common multiple of 3 and 5, and also of 3 and 10. Can you explain that?

Children discuss any patterns and relationships they notice.

What do you notice about all the numbers that are multiples of both 2 and 3?

What about the common multiples of 2 and 4? And 2 and 5?

What can you say about all the numbers that are common multiples of 3 and 4? And 3 and 5?

Assessing using and applying

- Children can identify some common multiples of two numbers to 10 (to the tenth multiple).
- Children can identify the common multiples of two numbers to 10 (to the tenth multiple) and can make generalisations about these. For example: "12, 24 and 36 are common multiples of 3 and 4, and they are also all multiples of 12."
- Children can identify the common multiples of two or more numbers, including those beyond the tenth multiple, and can make generalisations about these.

Supporting the challenge

- Children colour in the multiples of two numbers, using a different colour for each. This way, they can see that the common multiples clearly have both colours.
- Children concentrate on finding common multiples of the 2, 3, 4, 5 and 10 times tables.

Extending the challenge

- Children identify common multiples in three different times tables: for example, common multiples of 2, 3, 5.
- Children identify common multiples beyond the tenth multiple.

Brick wall problem

Challenge 12 **Using multiplication facts to solve a puzzle**

Using and applying

Representing

Represent a puzzle using diagrams

Reasoning

Identify and use patterns, relationships and properties of numbers; investigate a statement involving numbers and test it with examples

Maths content

Knowing and using number facts

- Derive and recall multiplication facts to 10×10

Calculating

- Develop and use written methods to record, support and explain multiplication of two-digit numbers by a one-digit number

Key vocabulary

multiplication, product

Resources

For each pair:

- RS15 (optional)
- Pencil and paper

How many different numbers are possible for the top brick?

Introducing the challenge

Draw a brick wall with a foundation of three bricks and write the numbers 1 to 5 to one side of the wall. Children choose three of the numbers and write them in the three bricks at the bottom of the wall.

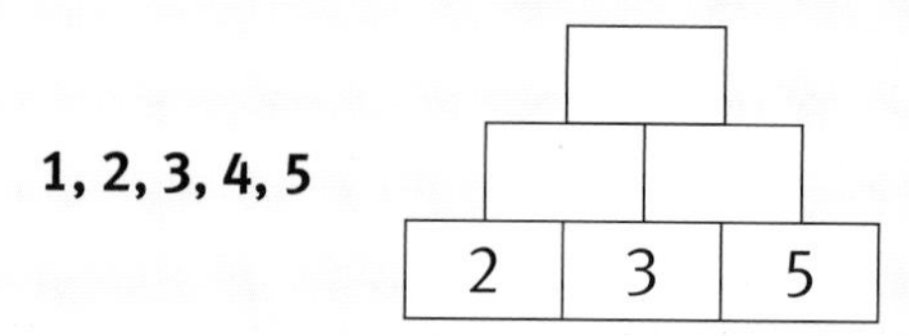

Show how this is a product wall. To work out the numbers in the bricks in the next layer up, multiply together adjacent pairs of numbers and write the product in the brick above. To find the number in the top brick, multiply together the two numbers in the middle layer. Work this out with the children, inviting answers to each calculation. Record on the bricks.

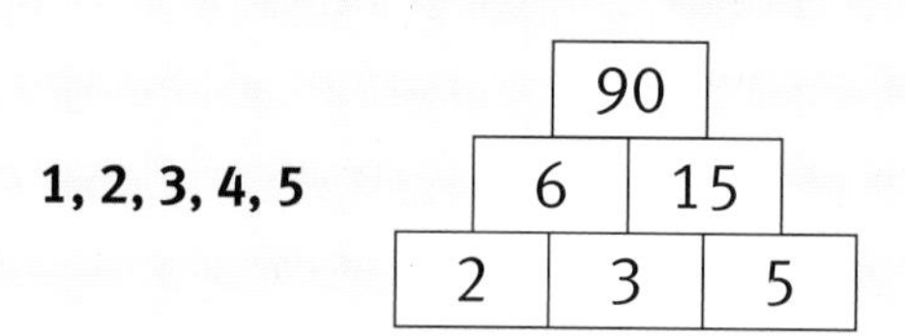

Repeat, starting with other numbers in the bottom three bricks.

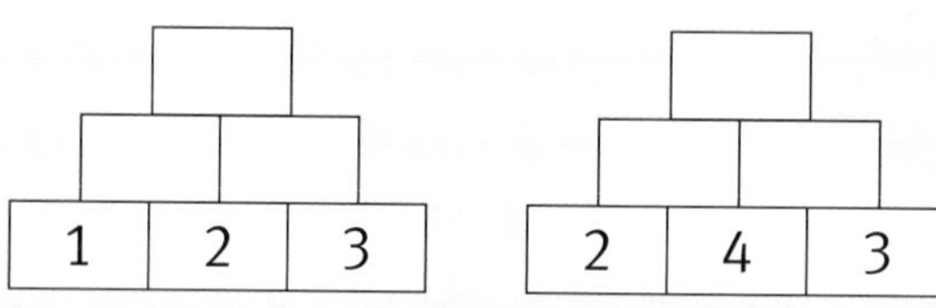

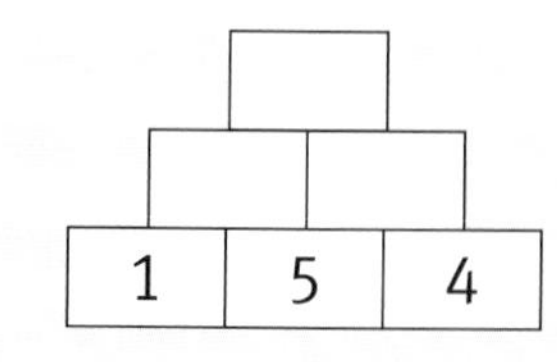

The challenge

Introduce the challenge to the children. They can use RS15 to record on. This time, they will start with a four-foundation brick wall, using the numbers 1, 1, 2 and 3 on the bottom layer. The challenge is to work out how many different numbers are possible for the top brick, using different combinations of 1, 1, 2 and 3 in the bottom row.

How do you know you have written all the different combinations?

Do you think the number on the top brick will be the same each time or different? Why?

How will you make the smallest number on top? The largest? Why does that work?

Drawing out using and applying

Invite individual pairs of children to comment on the number of different brick wall combinations there are (6, not counting mirror images) and the four different top brick products. Ask children to talk about the different strategies they used to discover the different brick wall combinations. One strategy is to consider the different centre numbers in the bottom four bricks, and then the different possible combinations for the other two numbers.

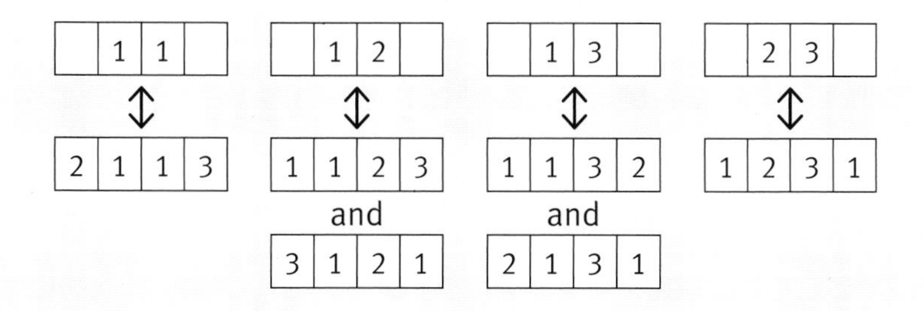

How many different combinations of brick walls are there?

How do you know you have found all the different combinations?

Did you use any strategies when you were working out how many different combinations of brick walls there were? What were they?

Did anyone use a different strategy?

What bottom row gives the smallest number on the top brick? The largest? How does that work?

What bottom rows give the same number on the top brick? Why?

Assessing using and applying

- Children can find the product in the top brick.
- Children can find different combinations of brick walls and their products.
- Children can work systematically to find all the different combinations of brick walls and their products.

Supporting the challenge

- Help children multiply a two-digit number by a one-digit number by doubling and doubling again.
- Children add (or find the difference between) pairs of adjacent numbers. How many different numbers are possible for the top brick?

Extending the challenge

- Children try different numbers on the bottom bricks.
- Children try brick walls of different sizes.

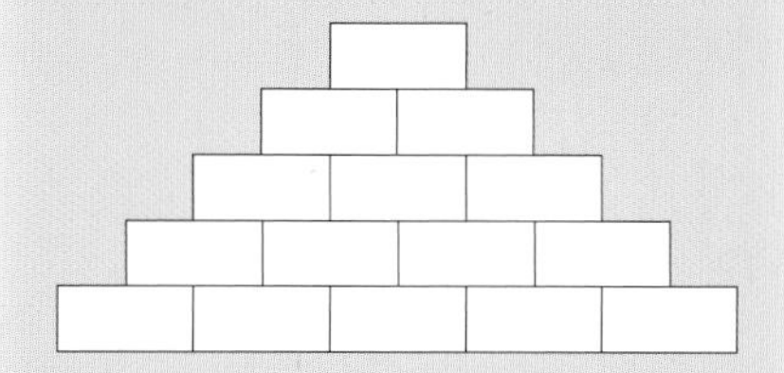

Folding a number square

Challenge 13

Using multiplication and division facts to solve a puzzle

Using and applying

Representing

Represent a puzzle using number sentences; use these to solve the problem

Communicating

Report solutions to puzzles, giving explanations and reasoning orally and in writing, using diagrams and symbols

Maths content

Knowing and using number facts

- Derive and recall multiplication facts to 10×10 and the corresponding division facts
- Use knowledge of rounding, number operations and inverses to estimate and check calculations

Key vocabulary

multiplication, division, factor, product

Resources

- Large demonstration 2×2 blank squares

For each pair:

- RS16
- RS Extra (2×2 squares) or squares of paper
- Scissors
- Pencil and paper

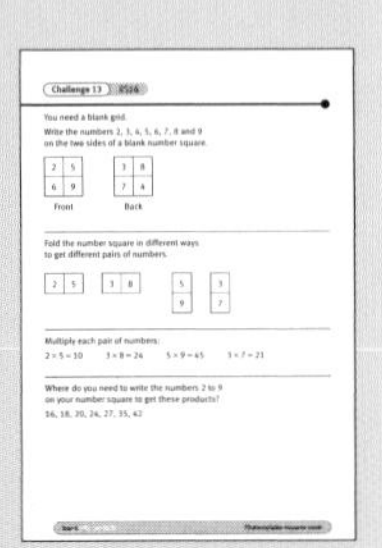

Can you work out where to put the numbers on the number square?

Introducing the challenge

Write the numbers 1, 2, 3, 4 on one side of the demonstration square and 5, 6, 7, 8 on the other.

Fold the square in different ways to reveal different pairs of numbers (both horizontally and vertically).

Multiply together each pair of numbers, inviting children to help you. Write these up as you go.

Invite children to help you work out systematically how many different products there are. There will be eight altogether, depending on how you arranged the numbers. One way is: 1×2, 1×4, 2×3, 4×3; 5×6, 5×8, 6×7, 8×7

Arrange the same numbers differently on another demonstration square to show different products. You might put 1, 3, 5, 7 one side and 2, 4, 6, 8 the other. Again, there will be eight products, depending on where you write them. For example: 1×3, 1×5, 3×7, 5×7;
2×4, 2×6, 6×8, 4×8

Pairs of children cut out 2×2 squares from RS Extra or fold squares of paper in four and write the numbers on the smaller squares. Write up the numbers 2, 3, 4, 5, 6, 7, 8, 9. Pairs use these numbers to write on the front and back of their squares in any way they like.

They find all the eight possible products for their arrangement of those numbers.

Once the children have done this, discuss how they worked out how many different combinations there were.

How many different calculations are there? What are they?

How do you know you have found all the different calculations? What did you do to make sure?

The challenge

The challenge is to find the arrangement of 2, 3, 4, 5, 6, 7, 8, 9 on the two sides of the 2×2 square that will give the products 16, 18, 20, 24, 24, 27, 35, 42.

Some pairs of children will rush into making their number square haphazardly. Others will use trial and improvement, while some may identify factors and common factors. Encourage the children to use a systematic way of working out the position of the eight numbers.

How do you think you might approach this problem?

What are you going to do first?

Can you work backward from the products? What numbers do you multiply together to get 16? How does that help?

How are you going to keep a record of what numbers you have tried?

What does this combination of numbers tell you?

Drawing out using and applying

Invite pairs of children to suggest the position of the eight numbers on the number square.

Has anyone else positioned the numbers in a different way?

Do the answers to the number square still work? Why is that?

Focus on how the children worked out the position of the numbers.

How did you work out where the eight numbers needed to go?

Who used a different method?

Did anyone use any other methods?

Which methods do you think work the best? Why?

Finally, ask a pair of children who made their own number square and worked out the position of their partner's numbers to talk about what they did (see *Supporting* and *Extending the challenge*).

Assessing using and applying

- Children can work haphazardly to identify the position of some numbers on the number square.
- Children can work systematically, using trial and improvement to identify the position of the numbers on the number square.
- Children can use their knowledge of factors to identify positions of numbers on the number square and then modify their answers, using trial and improvement.

Supporting the challenge

- Tell the children the position of two of the eight numbers.
- Children use the numbers 2 to 9 and add pairs of numbers together. They give their partner the answers, who then works out the position of the numbers.

Extending the challenge

- Children make a number square. They use eight numbers and multiply pairs of numbers. They tell their partner the eight numbers and the answers to the eight calculations. The other child then works out the position of the eight numbers.
- Children make a 3×3 number square. They write 1 to 9 on each side of the square, add the three numbers together and give their partner the answers to the 12 calculations. The other child then works out the position of the 18 numbers.

Front

2	5	3
7	1	8
4	6	9

Back

9	6	7
4	2	8
3	5	1

Doubling trouble

Doubling whole numbers and using mental and written methods of multiplication

Challenge 14

Using and applying

Reasoning

Identify and use patterns, relationships and properties of numbers; investigate a statement involving numbers and test it with examples

Communicating

Report solutions to problems, giving explanations and reasoning orally and in writing, using diagrams and symbols

If I double each number in a multiplication calculation, what happens to the answer?

Introducing the challenge

On the board, write a two-digit number less than 50, with the units digit greater than 5: for example, 38. Children double the number in their head or with jottings. Ask one or two children how they did it. Then demonstrate how you can partition the number into tens and units, double each separately and add the doubles.

38 = 30 + 8 **double 30 = 60** **double 8 = 16**
60 + 16 = 76 **double 38 = 76**

Children may have partitioned the number in other ways: for example, by thinking of 38 as 35 + 3 and doubling each of these.

Repeat this for two or three other numbers.

Give each child five paper cards. In pairs, the children write a number under 50 on each of their cards in the same colour (for example, blue). Pairs pool their cards and together figure out the double of each number, writing the answer on the back in a different colour (for example, red).

Children spread out the paper cards, with all the original numbers (blue in this example) face up. They take turns to select a number and say what is on the back of it. They check by turning it over so only they can see. If they are right, children show the answer to their partner and keep it. If they were wrong, they put the number back on the table. They continue until they have collected all 10 cards.

Maths content

Knowing and using number facts

- Identify the doubles of two-digit numbers; use these to calculate other doubles

Calculating

- Develop and use written methods to record, support and explain multiplication of two-digit numbers by a one-digit number

Key vocabulary

double, twice, multiplication

Resources

- Pieces of paper about the size of playing cards (five for each child)
- Pens in two different colours
- RS17 (optional, for each pair)

For *Supporting the challenge*:

- Interlocking cubes

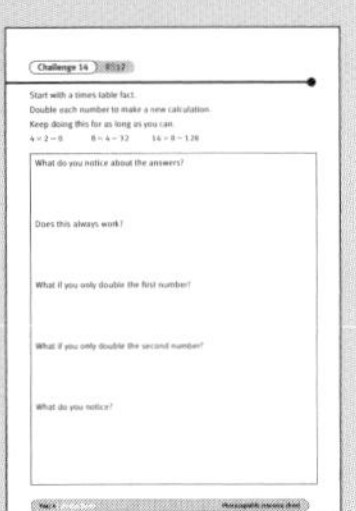

The challenge

The challenge is to start with a simple multiplication such as 4×2, then to double both numbers to 8×4, 16×8, and so on, and find the answers to the calculation each time. Children need to look for patterns between the answers (each answer is 4 times, or double and double again, the previous answer).

What do you notice about each of your answers?

Does this always work?

Once the children have tried this, using several known multiplication facts, ask them to answer the 'What if?' questions.

What if you only double the first number?

What if you only double the second number?

What do you notice?

Children can record on RS17.

Drawing out using and applying

Provide an opportunity for pairs of children to explain to the rest of the class what they noticed. Ask children to offer a generalisation.

What happens to the answer when you double each of the numbers in a multiplication calculation?

Does this always work?

Who can tell me a statement about the rule we have discovered?

What else can you say?

Could you use this rule to make any other statements? How could we check if the statement is correct?

Assessing using and applying

- Children can double two-digit numbers and multiply numbers, but do not notice any patterns.
- Children can make a generalisation such as: "When you double each of the numbers in the multiplication calculation, the new answer is always four times the previous answer."
- Children can make a conjecture based on a generalisation such as: "When you halve one of the numbers in the multiplication calculation, the new answer is always half of the previous answer."

Supporting the challenge

- Give children simpler multiplications: 2×1, 2×3, 2×4, 3×4
- Children use cubes or similar apparatus to model the doubling.

Extending the challenge

- Does this rule apply to decimal numbers?
- What if you started with an addition number fact and doubled each of the numbers?

Dining out

Challenge 15

Solving problems involving money

Using and applying

Solving problems

Solve problems involving money; choose and carry out appropriate calculations, using calculator methods where appropriate

Enquiring

Suggest a line of enquiry and the strategy needed to follow it; collect, organise and interpret selected information to find answers

Maths content

Calculating

- Refine and use efficient written methods to add and subtract two-digit and three-digit whole numbers and £.p
- Use a calculator to carry out one-step and two-step calculations involving all four operations; interpret the display correctly in the context of money

Key vocabulary

addition, total, money, price, cost, pounds, pence, share

Resources

For each pair:

- RS18
- Large sheet of paper and marker

For *Supporting the challenge*:

- Calculator

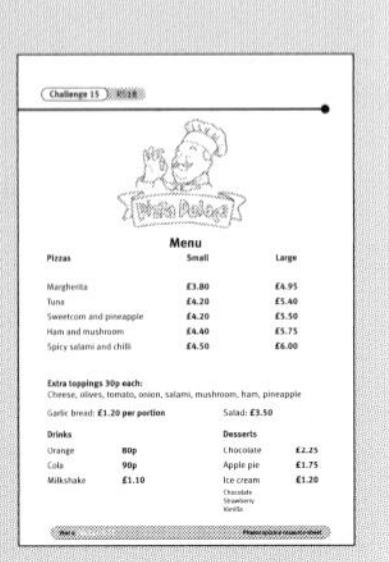

Challenge 15 | RS18

Menu

Pizzas	Small	Large
Margherita	£3.80	£4.95
Tuna	£4.20	£5.40
Sweetcorn and pineapple	£4.20	£5.50
Ham and mushroom	£4.40	£5.75
Spicy salami and chilli	£4.50	£6.00

Extra toppings 30p each:
Cheese, olives, tomato, onion, salami, mushroom, ham, pineapple

Garlic bread: **£1.20 per portion** Salad: **£3.50**

Drinks	
Orange	80p
Cola	90p
Milkshake	£1.10

Desserts	
Chocolate	£2.25
Apple pie	£1.75
Ice cream (Chocolate, Strawberry, Vanilla)	£1.20

What will you and your friends have to eat at Pepe's Pizza Palace?

Introducing the challenge

Discuss going out for a meal and talk about the children's experiences and preferences. Compare their knowledge of prices.

What food do you like to eat? To drink?

Who has eaten in a restaurant? What type of food was it?

How much might ... cost?

The challenge

Provide each pair with a copy of RS18, a large sheet of paper and a marker. Read through and discuss the resource sheet with the class. Children decide how many friends to plan a meal for, but you may prefer to direct individual pairs of children to work out a menu for specific numbers to ensure an appropriate challenge.

What will you and your friends have to eat at Pepe's Pizza Palace?

How do you know who has ordered what?

Have people ordered drinks? What about desserts?

How are you keeping a record of what you are doing?

How are you going to share the bill between you?

What calculations does that involve?

Tell the children that the large sheet of paper is for them to plan their meal and to work out how much each person will pay. Explain that some pairs of children are going to show their sheet to the class and use this to talk through what they did.

Drawing out using and applying

Provide an opportunity for pairs of children to report back to the rest of the class, using their large sheet of paper as a crib. Discuss the various methods children used to record what food individual people ordered, and how they worked out how much each person would need to pay. Draw attention to those pairs who itemised the food and costings in the form of a restaurant bill. Ask the class to comment on the recording and calculating methods used by individual pairs of children.

How many people was your menu for? What was the total bill for the meal? How did you work it out?

How did you work out how much each person had to pay?

What are the different ways you could work out how much each person had to pay? Was this fair? Why? Why not?

How did you record what each person ordered? What was good about what you did?

What could you have done differently to make it easier to know who had what to eat? ... what the total bill was? ... how much each person needed to pay?

How else could you have recorded the meal and the costings?

Assessing using and applying

- Children can organise the problem systematically and clearly, showing by their recording what the separate orders are and what the total bill is.
- Children can challenge themselves by choosing a number of guests, appropriate for their own skills, to calculate with.
- Children can find solutions to solve the last part of their own problem, sharing out the bill between the participants of the meal.

Supporting the challenge

- Children plan a meal for a family of four.
- Children use a calculator to help with calculating the bill and then sharing it between the group.

Extending the challenge

- Children include a 10% service charge in the bill.
- Children invent a class take-away café, making and working out menus, order forms, bills, delivery times, and so on.

How much paper?

Challenge 16

Estimating, counting and calculating

Using and applying

Enquiring

Suggest a line of enquiry and the strategy needed to follow it; collect, organise and interpret selected information to find answers

Communicating

Report solutions to problems, giving explanations and reasoning orally and in writing, using diagrams

Maths content

Calculating

- Develop and use written methods to record, support and explain multiplication
- Use knowledge of rounding, number operations and inverses to estimate and check calculations

Key vocabulary

estimation, approximation, calculate, multiplication

Resources

- RS19 (optional, for each pair)
- Calculators (optional)

For *Supporting the challenge*:

- Calculator

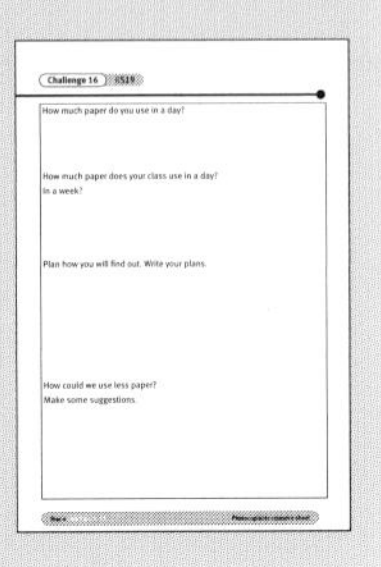

How much paper do you use?

Introducing the challenge

Brainstorm with the class the use of paper in the classroom and ask for ideas about how to investigate how much paper is used in one day. Strategies for working out individual use will differ from class use, and days may vary according to the class's timetable – these are factors the children need to take into consideration. They need to decide how to quantify paper: loose paper is measured in reams (500 sheets) and bound paper in exercise books by a set number of pages (usually a multiple of eight). The class will also make use of a variety of assorted papers for artwork and for all kinds of incidental use.

How much paper do you think you each use in one day? Talk to the person next to you.

When throughout the day do we write things down? Draw things? Use paper for other reasons?

What types of paper do we use?

Is this the same every day of the week?

The challenge

Children work in pairs to plan how they would find out how much paper each person uses in a day, how much the whole class uses in a day, and how much this is in a week. They can use RS19 to record. They need to decide what information to collect and how to collect it. Have calculators available for children who are less confident working with larger numbers. Encourage the children to make estimations and approximations.

What things are you going to want to find out?

How are you going to go about finding these out?

What is different about a piece of photocopied paper and a sheet of paper in your exercise book? What about a piece of art paper? Poster paper?

From the amount you use, can you guess how much the class uses?

How could you make your estimate more accurate?

Compare your estimate with someone else's. How different is it? Is this a significant difference? Why do you think it is different?

Explain to the children that when they have finished working out how much paper they use in a day/week, they are to discuss with their partner how they could use less paper and be prepared to offer some suggestions at the end of the lesson.

Drawing out using and applying

Children present their results to the class, showing what calculations they made and how they arrived at their final estimates. Ask the children to reflect on the differences and to decide to what extent the differences matter.

How did you work out how much paper you use in a day?

What about for the whole class?

What calculations did you make?

Do you think that method is accurate? Might it be accurate enough? Can you explain why?

Does a difference of 10 matter in 1000?

What about a difference of 100?

When does a difference start to matter?

Finally, ask the children to suggest ways of using less paper.

How could each of us use less paper each day/each week?

What could we do instead? What could we use?

Would these suggestions work for the class as a whole? Are there other things that we could do as a class to use less paper?

Why should we want to be using less paper?

Assessing using and applying

- Children can suggest ways of tackling the problem and offer a workable plan.
- Children can estimate by calculating, using approximation and rounding.
- Children can present their findings to the class clearly and justify their results.

Supporting the challenge

- As the challenge involves calculating with large numbers, children can use a calculator.
- Some children may find the open-ended nature of the investigation daunting. If so, ask them leading questions which will assist them in making and carrying out a plan to find out how much paper they use.

Extending the challenge

- Investigate measurements of quantities of paper and sizes of paper.
- Find out how much paper you can get from one tree and compare with the class's use.

Tyler's cakes

Challenge 17

Solving numerical problems

Using and applying

Solving problems

Solve one-step and two-step problems involving numbers; choose and carry out appropriate calculations, using calculator methods where appropriate

Communicating

Report solutions to problems, giving explanations and reasoning orally and in writing, using diagrams

Maths content

Calculating

- Add, subtract, multiply and divide mentally
- Refine and use efficient written methods to add, subtract, multiply and divide

Key vocabulary

addition, subtraction, multiplication, division, problem, calculation

Resources

For each pair:

- RS20
- Pencil and paper

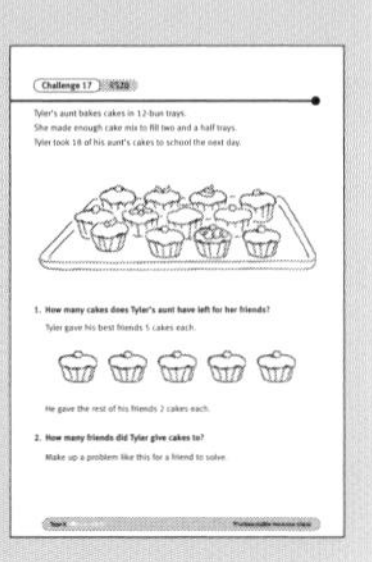

Challenge 17 RS20

Tyler's aunt bakes cakes in 12-bun trays.
She made enough cake mix to fill two and a half trays.
Tyler took 18 of his aunt's cakes to school the next day.

1. How many cakes does Tyler's aunt have left for her friends?

Tyler gave his best friends 5 cakes each.

He gave the rest of his friends 2 cakes each.

2. How many friends did Tyler give cakes to?

Make up a problem like this for a friend to solve.

How many friends did Tyler give cakes to?

Introducing the challenge

Pose this problem to the children:

I have 6 cakes. I give 2 cakes each to some friends and 1 cake to some others. How many friends could I give cakes to?

In pairs, children discuss how to solve this problem. Write up different solutions on the board.

The challenge

Provide each pair with a copy of RS20 and talk it through. Children work together on the problems. If they have difficulty knowing where to start, ask them:

What information does the problem give you?

Can you make a drawing or write something that will help you sort out the information?

Would it help to work backward?

What if you substituted the numbers in the problem with smaller numbers?

What if you just guessed and tried it out?

Children record how they solved the problems, using numbers and symbols. As they work, ask them questions that might help them when writing their own problems:

How many are left after Tyler gives away 7 cakes?

How many friends can have 3 cakes each?

How many cakes are there in 4 similar trays?

Tyler's aunt bakes 40 cakes. How many trays does she use to bake all of the cakes?

Tyler's aunt decides to sell a full tray of cakes, at 24p a cake. How much money would she make if she sold three quarters of the cakes?

When the children have answered the problems, ask them to make their own problems and offer them to the rest of the class to solve.

Drawing out using and applying

Ask the children to explain to the class how they tackled each of the problems and solved it. In particular, ask children who had difficulty in starting to describe what strategies they used.

Invite those children who made up their own word problems to present these to the rest of the class. The remainder of the class use their whiteboards to work out and display the answer. Again, encourage the children to discuss the strategies they used for working out the answer.

What is your answer?

How did you work it out?

Did anyone get a different answer?

Did anyone work out the answer in a different way?

Assessing using and applying

- Children can reason through a problem.
- Children can try different approaches if they are finding it difficult, such as working backward and drawing pictures.
- Children can explain their strategies for solving the problem.

Supporting the challenge

- Provide the children with a copy of page 11, which provides possible problem-solving strategies.
- Children write their own one-step word problems.

Extending the challenge

- Children write two-step or multi-step word problems.
- Children make use of real-life problems that are complex and need unravelling.

A fraction of your money

Challenge 18 **Finding fractions of money**

Using and applying

Solving problems

Solve problems involving money; choose and carry out appropriate calculations

Communicating

Report solutions to problems, giving explanations and reasoning orally

Maths content

Knowing and using number facts

- Use knowledge of rounding, number operations and inverses to estimate and check calculations

Calculating

- Find fractions of quantities

Key vocabulary

fraction, amount, estimate, multiplication, division, multiples, factors

Resources

For each pair:

- RS21
- Selection of coins

Which amounts of money make good fractions?

Introducing the challenge

Invite a child to the front. Give them 20p, then ask for half of this money back (they can change the 20p for smaller coins). Swap roles: this time, the child gives you 15p. Invite the class to work in pairs to help the child at the front decide what fraction they might ask for back.

Repeat this with different amounts.

What is half of £1.20? What is one quarter of 60p?

How did you work it out?

What is two thirds of 90p? What about three fifths of 90p?

What do you need to work out first?

The challenge

Provide each pair with a copy of RS21 and a selection of coins. Read through the resource sheet with the children and discuss the game with the class. The scoring is designed so that the children setting the problem have to think it through mentally first to make sure that it works, but also not to make it too easy for the other player.

Briefly begin to play the game, you against the rest of the class. Model a variety of good (and poor) coins to choose and fractions to ask.

As the children work, make sure that they are really challenging each other.

What amount of money have you chosen?

What fraction are you going to ask for?

Is this a good amount/fraction? Why?

Drawing out using and applying

Invite pairs of children to share their results of the game with the whole class. Ask children to report back on the hardest fraction they could think of that worked, and the hardest fraction they had to solve. Discuss methods of calculating the fractions in their heads and ways of checking results by 'working backward'.

Which amounts of money are good to give to your partner? Why?

Which are not so good? Why?

What strategies did you use to make sure your partner could always give back the fraction?

Assessing using and applying

- Children can invent a problem and work out the answer in their heads.
- Children can solve a problem posed by someone else.
- Children can check each other's solutions for accuracy.

Supporting the challenge

- Suggest to children that they work backward. For example: "Start with 5p and multiply this by 4 to get 20p. You now know that 5p is a quarter of 20p. Find a 20p coin and ask: What is one quarter of 20p?"
- Encourage the children to ask only unitary fractions of amounts: for example, $\frac{1}{2}$ of 70p, $\frac{1}{4}$ of 40p and $\frac{1}{3}$ of 18p.

Extending the challenge

- Children invent problems in other areas of mathematics for their friends to solve and prepare solutions to the problems.
- Children develop explicit methods for checking calculations, such as reversing the operations, estimating for reasonableness, and so on.

No more sixes

Challenge 19

Understanding the relationship between the four operations

Using and applying

Solving problems

Solve problems involving numbers; choose and carry out appropriate calculations, using calculator methods where appropriate

Reasoning

Identify and use patterns, relationships and properties of numbers

Maths content

Knowing and using number facts

- Use knowledge of rounding, number operations and inverses to estimate and check calculations

Calculating

- Use a calculator to carry out calculations involving all four operations; correct mistaken entries and interpret the display correctly

Key vocabulary

addition, subtraction, multiplication, division, calculator

Resources

- Demonstration calculator
- Small sticky labels (optional)

For each pair:

- Calculator
- RS22
- Pencil and paper

Is the calculator any use, even though one number key is broken?

Introducing the challenge

Display the demonstration calculator. Ask the children to imagine that the '6' key on the calculator is broken – nothing happens to the screen if they press the '6' (you might want to put a small sticky label over the '6' key). However, you can still show a '6' on the calculator display. (If the calculators being used have memories and the children know how to use these, ask them not to use the memory function.)

Ask pairs of children to decide how they could get 666 to show on the calculator display without using the '6' key. Emphasise that they need to decide which keys they are going to press before using the calculator.

How would you get the calculator display to show 666 without pressing the '6' key?

What calculation might you use?

Is there another calculation you could key in that doesn't use the number 6?

Discuss with the children the different methods they thought of.

Write the following calculation on the board: 16 + 6. Explain to the children that they shouldn't need to use a calculator to work out the answer, but given that the '6' key is broken, how would they go about answering this calculation? Discuss the different methods suggested by the children.

The challenge

Distribute the calculators to pairs of children. Explain that pairs will work on the problem involving the calculator and that there are two particular rules they need to stick to. Firstly, they must take it in turns to operate the calculator, and secondly, they must agree on what they key into the calculator before either partner presses a key.

Give each pair a copy of RS22. The challenge is to use the calculator to find the answers to the calculations, still imagining that the '6' key is broken.

Children who finish early could make up some similar problems to challenge their friends with.

Drawing out using and applying

Discuss the various strategies children used to get round the broken '6' key.

How did you work out the answer to 48 multiplied by 6?

What did you do to get round not being able to use the number 6 on the calculator?

Did anyone use a different strategy?

Invite children who made up similar problems to present these to the rest of the class.

Assessing using and applying

- Children can solve 'broken 6' problems involving one 6 and addition or subtraction.
- Children can solve 'broken 6' problems involving more than one 6 and addition or subtraction.
- Children can solve 'broken 6' problems involving all four operations.

Supporting the challenge

- Ask the children to solve simpler calculations involving the number 6. For example:
  TU ± U (36 ± 8)
  TU ± TU (48 ± 26)
  U × U (6 × 7)
  TU ÷ U (86 ÷ 4)
- Remind the children of how to partition numbers in different ways and use one of the calculations to demonstrate how this might be a useful strategy in overcoming the broken '6' key.

Extending the challenge

- Suppose the '2' and '6' keys on the calculator are broken.
- Suppose the multiplication key is broken.

Tables for Bingo

Challenge 20

Working with written multiplication and division

Using and applying

Reasoning

Identify and use patterns, relationships and properties of numbers

Communicating

Report solutions to problems, giving explanations and reasoning orally and in writing, using diagrams and symbols

Maths content

Knowing and using number facts

- Use knowledge of rounding, number operations and inverses to estimate and check calculations

Calculating

- Develop and use written methods to record, support and explain multiplication and division

Key vocabulary

multiplication, division, pattern, relationship

Resources

- RS23 (for each child)
- Calculating equipment such as a 100-grid, number line or calculator

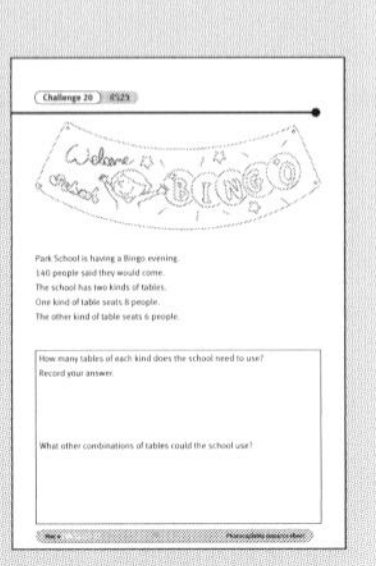

How many tables of each kind did the school use?

Introducing the challenge

Pose this problem to the children:

A school is holding a Bingo evening. 30 people are coming. The school has two kinds of tables, three-seaters and five-seaters. How many tables of each kind does the school need to use to seat all 30 people?

Pairs discuss how to solve this problem.

Invite a few children to explain how they solved it.

Discuss the variety of solutions and methods with the children.

Is there another solution to the problem?

Could there be another combination of three-seater and five-seater tables?

Are there any more?

How can you be sure that you have found all the possible combinations?

Make sure that the children realise that there is more than one solution to the problem.

6 five-seaters

10 three-seaters

3 five-seaters and 5 three-seaters

The challenge

Each child has a copy of RS23 and appropriate calculating equipment. Encourage the children to draw a diagram of their solution.

How many tables of each kind did the school use?

Is this the only possibility? Is there another one? Are there any more?

Once children have answered the problem, they compare their solution with a partner and discuss and compare methods. Explain to them that they may be asked to report back to the whole class on their partner's solution and their method of working towards the end of the lesson.

Can you explain to me your partner's solution?

Explain to me how your partner worked out their solution(s)?

Drawing out using and applying

Ask one child to share and compare their partner's solutions and to explain and describe their methods of working. Repeat for other pairs of children. There will be several different solutions.

What solution did you come up with?

Is there another solution?

How many different solutions are there?

How can we be sure that we have found them all?

What are good methods to start solving the problem?

How did you keep a record of what you did?

Assessing using and applying

- Children can explain their methods, including how they started solving the problem.
- Children can work systematically to find all the possible solutions and describe their system: "I started with one six to see if the eights could make up the rest, then two sixes, and so on."
- Children can make a visual representation of all of the solutions.

Supporting the challenge

- Provide the children with a variety of calculating apparatus such as a 100-grid, number lines and a calculator.
- Change the numbers in the problem to make calculating easier: for example, 68 people and 4-seater and 6-seater tables.

  2 × 4-seaters + 10 × 6-seaters

  5 × 4-seaters + 8 × 6-seaters

  8 × 4-seaters + 6 × 6-seaters

  11 × 4-seaters + 4 × 6-seaters

  14 × 4-seaters + 3 × 6-seaters

Extending the challenge

- Children make up some more problems of the same kind and discuss ways of inventing the problems.
- Children investigate the numbers of people they cannot seat with 6-seater and 8-seater tables without leaving a space on the table. They investigate other kinds of tables.

Challenge 21

Phone numbers

Using mental and written methods for addition to solve a puzzle

Using and applying

Solving problems

Solve problems involving numbers; choose and carry out appropriate calculations

Representing

Represent a puzzle using number sentences; use these to solve the problem; present and interpret the solution in the context of the problem

Maths content

Calculating

- Add or subtract mentally pairs of two-digit whole numbers
- Refine and use efficient written methods to add and subtract two-digit and three-digit whole numbers

Key vocabulary

number, digit, place value, addition, subtraction, multiplication, division

Resources

For each child:

- RS24 (optional)

For *Supporting the challenge*:

- Number lines

What different totals can you make using all the digits in your telephone number?

Introducing the challenge

On the board, write an eight-digit telephone number such as 16031929. Ask the children to suggest different ways in which these eight digits can be separated into one-digit or two-digit numbers. Explain that they need to use all the digits and that they cannot change the order of the digits.

16 03 19 29   **1 60 3 19 29**

16 0 3 19 29   **16 0 3 1 92 9**

The challenge is to add each string of numbers.

Work with the children to find the totals. For example:

$16 + 03 + 19 + 29 = 67$

$1 + 60 + 3 + 19 + 29 = 112$

$16 + 0 + 3 + 19 + 29 = 67$

$16 + 0 + 3 + 1 + 92 + 9 = 121$

The challenge

The challenge is to do the same with their own eight-digit telephone number (or children can make up one). They also have to work out what is the largest answer they can get by separating the digits into one-digit or two-digit numbers and finding the total.

Children can record on RS24 if this is helpful.

Drawing out using and applying

Ask individual children for the telephone number they used and one or two of their calculations. Discuss with the class how different children worked through the different combinations of one-digit and two-digit numbers.

What are some of the different totals you can make using the digits that make up your telephone number?

What is the largest answer you can get by separating the digits into one-digit or two-digit numbers and finding the total? How do you know that is the largest possible number?

Did anyone make a larger number than this? How come it is larger?

What is the smallest answer you can get by separating the digits into one-digit or two-digit numbers and finding the total?

Assessing using and applying

- Children can separate the digits in their telephone number to make different one-digit or two-digit numbers and add them together.
- Children can say the largest and smallest answers they can get by separating the digits into one-digit or two-digit numbers and finding the total.
- Children can make a generalisation about what makes the smallest and largest totals. For example: "The smallest answer can be made by arranging the digits into one-digit numbers and adding them together."

Supporting the challenge

- Children use a four-digit number such as the year of their birth and separate this into one-digit and two-digit numbers.
- Children use number lines to model the calculations.

Extending the challenge

- Children separate the digits into one-digit, two-digit or three-digit numbers and find the total.
- What if you separated the digits to make subtraction calculations? However, you must make sure that your answers are always positive numbers.

Would you rather?

Challenge 22

Finding fractions of numbers

Using and applying

Solving problems

Solve problems involving numbers; choose and carry out appropriate calculations

Communicating

Report solutions to puzzles, giving explanations and reasoning orally and in writing, using diagrams and symbols

Maths content

Knowing and using number facts

- Derive and recall multiplication facts to 10×10 and the corresponding division facts

Calculating

- Find fractions of numbers

Key vocabulary

number, fraction, multiplication, division

Resources

For each pair:

- Individual whiteboard and pen
- RS25

For *Supporting the challenge*:

- Multiplication square (see RS Extra for Challenge 11)

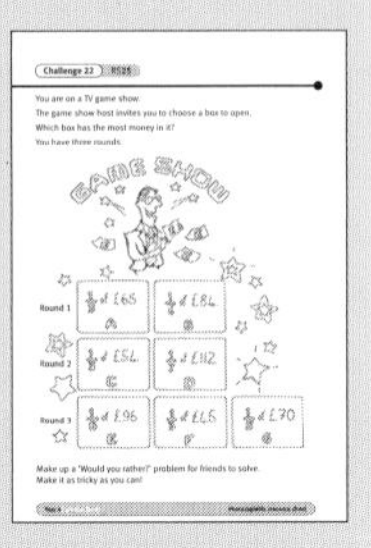

Which would you rather have: a third of £30 or a quarter of £60? Why?

Introducing the challenge

Pose the following question to the children:

Which would you rather have: a third of £30 or a quarter of £60?
(a third of £30 = £10; a quarter of £60 = £15)

Ask the children to turn to their partner and briefly discuss the question. Tell them to do any working out on their individual whiteboards.

Ask various pairs of children for their preference, justifying their decision.

Which would you rather have? Why?

How did you work it out?

Did anyone work it out another way?

Repeat the above, posing the following question to the children:

Which would you rather have: a quarter of £56 or a ninth of £108?
(a quarter of £56 = £14; a ninth of £108 = £12)

Discuss children's preferences and reasoning.

The challenge

Provide each pair of children with a copy of RS25. Briefly read through and discuss the problems with the class. Make sure the children realise that the first two problems involve comparing the amount of money in the two boxes, and that the third problem involves comparing the amount of money in three boxes.

Drawing out using and applying

Referring to each of the problems in turn, ask various pairs of children for their preference, justifying their decision.

Which would you rather have: a fifth of £65 or a sixth of £84? Why?

How did you work it out?

Did anyone work it out another way?

What if it was half of each amount? What would you choose then?

Finally, invite pairs of children who made their own 'Would you rather?' problem to share it with the rest of the class. Allow sufficient time for the class to solve the problems before asking for answers and discussing them.

What was the answer to Manesh and Stacey's problem?

Why was it a good one?

What made it tricky to answer?

Assessing using and applying

- Children can work out the answer, using trial-and-improvement techniques.
- Children can estimate the answer to the problem, using known multiplication and division facts.
- Children can create their own 'Would you rather?' problem where the difference between amounts of money is small: for example, £2 or less.

Supporting the challenge

- Ask 'Would you rather?' problems that involve the children calculating with unitary fractions that involve multiplication and division tables for which they have instant recall: for example, $\frac{1}{2}$, $\frac{1}{3}$, $\frac{1}{4}$, $\frac{1}{5}$ and $\frac{1}{10}$.
- Children use a multiplication square and work out the fraction of amounts.

Extending the challenge

- Ask 'Would you rather?' problems that involve the children calculating with non-unitary fractions: for example, "Which would you prefer to receive as pocket money: five sixths of £24 or three sevenths of £49? Why?"
- Invite children to invent their own 'Would you rather?' fraction problems. They should make them difficult to work out.

Describe it

Challenge 23

Using mathematical language for shape and position

Using and applying

Reasoning

Identify and use patterns, relationships and properties of shapes

Communicating

Report solutions to puzzles and problems, giving explanations and reasoning orally and in writing, using diagrams

Maths content

Understanding shape

- Draw polygons by identifying their properties
- Recognise horizontal and vertical lines; use the vocabulary of position

Key vocabulary

2D shape, regular, irregular, pattern, circle, triangle, isosceles triangle, right-angled triangle, rectangle, square, parallelogram, hexagon, straight, curved, side, length, parallel, perpendicular, horizontal, vertical

Resources

- RS26 (made into an OHT or an IWB image)
- RS27 (for each child)

For each pair:

- Pencil and paper
- Coloured pencils

For *Supporting the challenge*:

- Building blocks
- Interlocking cubes

Can you give good instructions so that your partner can recreate your design?

Introducing the challenge

Explain to the children that they will work in pairs (Child A and Child B). Give out these instructions:

- Child B, close your eyes and don't look at the board. (Display an image on the board, RS26.)
- Child A, look at the image. Don't write or draw anything. Try to remember the image. (Remove the image from the board after a few minutes.)

Child A, describe the image to Child B. Child B, draw what Child A is describing. Child A must not draw anything. Child B, you can ask Child A questions to help you draw the image. (Give children enough time to describe and draw the image.)

Once Child B has made an attempt at drawing the image, display the image on the board to the whole class. Allow the children time to discuss and compare the two images.

Child A, what did you find difficult about this task?

Child B, what did you find difficult about this task?

What did you need to use to make the task easier?

What were useful words for Child A to use when describing the image?

What were good questions for Child B to ask?

The challenge

Give out RS27 to each child and explain the challenge. Emphasise that Partner B in each pair should not be able to see what Partner A has drawn.

Leave the children to draw their own designs and recreate each other's.

If time allows, children repeat the activity, drawing their new designs on the back of the sheet.

Drawing out using and applying

Discuss the language that the children used.

Which words or phrases were particularly useful?

What made the drawing of your partner's design easier?

When was it hard to know what to draw?

Assessing using and applying

- Children can use everyday language and describe their design in terms of what it looks like.
- Children can accurately name shapes. For example: "This is an octagon."
- Children can use the language of mathematics accurately to describe shape and position. For example: "Parallel to the other side."

Supporting the challenge

- Provide pairs of children with building blocks identical in both shape and colour. One child secretly makes a shape using their blocks and describes it to their partner who has to use their blocks to make the same shape. Children then swap roles.
- Children make shapes using interlocking cubes.

Extending the challenge

- Pairs of children decide on their own collection of shapes to work with.
- Each child secretly makes a design using new shapes that their partner has not yet seen.

Challenge 24

Designing a playground

Investigating the 2D representation of 3D space in proportion

Using and applying

Enquiring

Suggest a line of enquiry and the strategy needed to follow it; collect, organise and interpret selected information to find answers

Communicating

Report solutions to problems, giving explanations and reasoning orally and in writing, using diagrams and symbols

Maths content

Understanding shape

- Visualise 3D objects from 2D drawings

Measuring

- Choose and use standard units and their abbreviations

Key vocabulary

space, distance, location, length, centimetre (cm)

Resources

- Maps and plans, architect's tape, metre stick, trundle wheel, catalogues with playground equipment, health-and-safety guidelines (for each group)
- RS28 (optional), cm squared paper, ruler, coloured pencils (for each pair)

For *Supporting the challenge*:

- Plan of the school playground

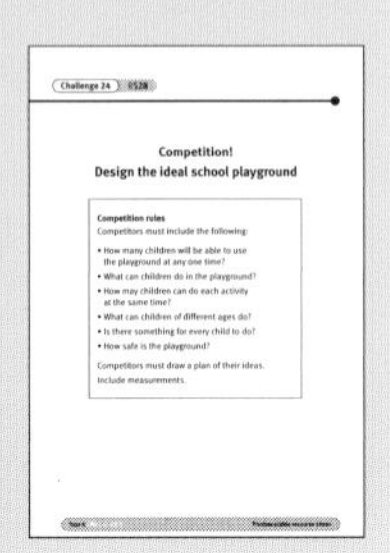

Challenge 24 | RS28

Competition!
Design the ideal school playground

Competition rules
Competitors must include the following:

- How many children will be able to use the playground at any one time?
- What can children do in the playground?
- How may children can do each activity at the same time?
- What can children of different ages do?
- Is there something for every child to do?
- How safe is the playground?

Competitors must draw a plan of their ideas.
Include measurements.

What would the ideal school playground be like?

Introducing the challenge

Brainstorm ideas about a school playground with the children.

What would the ideal school playground be like?

How many children could use it at a time?

Would it have something for children of different ages and with different interests?

What are the safety considerations?

Encourage the children to develop clear sets of guidelines for a playground: there could be no limit to cost, but it would have to be safe and would/would not (depending on the possibility of expansion) have to fit into the existing playground space in the school. When the children have agreed on these guidelines, they can work in pairs on the activity, using RS28 if they wish.

The challenge

Children need to research this activity. They need to measure the school playground, find out how many children there are in the school, browse through catalogues to see what equipment is available, and so on.

They also need to draw a plan of their ideal playground. Although they may not draw plans to scale at this stage, they should think about the proportions of the items they draw on it and find out, for example, how many benches would fit along the school wall and what size they should be on the plan.

How far does the climbing frame need to be from the door? Why?

Where will the balls in this area go?

How many children can sit in this area at the one time? Is that enough?

Where are the exits and entrances to the playground? How will you make them safe?

Drawing out using and applying

Invite pairs of children to share their results with the whole class. Ask the children to prepare formal presentations of their maps, plans and ideas to show to the class and expect to get questions from others about safety considerations and good use of space.

What do you like about Josh and Alex's design?

Is their plan clear about the amount of space different areas take up?

Would you like to play in this playground? Why?

How could their playground be improved?

How could their plan be improved?

Assessing using and applying

- Children can consider all the issues involved in designing a playground and take them into account in designing their ideal playground.
- Children can draw plans in reasonable proportion, with some idea of relative sizes of places for objects in their ideal playground.
- Children can explain and justify their design in terms of space and safety and question others in a reasoned way about their designs.

Supporting the challenge

- Provide the children with a simple plan of the existing school playground and ask them to improve it.
- Children design the ideal classroom.

Extending the challenge

- Children design the playground, working to a fixed budget.
- Children design and carry out a survey in the school to find out what improvements need to be made to the existing school playground.

Ideal gnome home

Challenge 25

Recognising features and properties of 3D shapes

Using and applying

Reasoning

Identify and use patterns, relationships and properties of shapes; investigate a statement and test it with examples

Communicating

Report solutions to puzzles and problems, giving explanations and reasoning orally and in writing, using diagrams and symbols

Maths content

Understanding shape

- Visualise 3D objects from 2D drawings
- Identify reflective symmetry

Key vocabulary

cube, symmetrical, combination, generalisation, statement

Resources

For each pair:

- RS29 (optional)
- Interlocking cubes

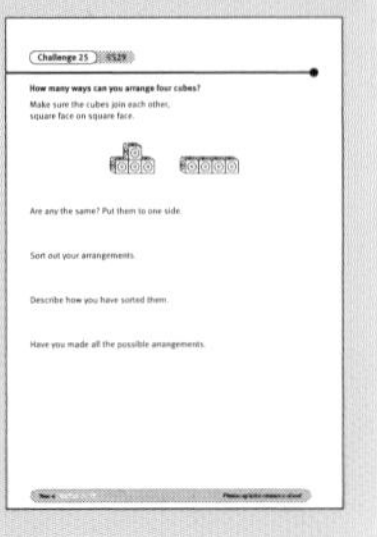

Use interlocking cubes to design some gnome homes.

Introducing the challenge

Provide each pair with a pile of interlocking cubes. Present the problem that gnomes live in houses with three rooms. Each cube represents a room.

Children make as many different houses as they can, using three interlocking cubes. Working individually first, then in pairs, children compare what they have done, discard duplicates and collect a complete set between them.

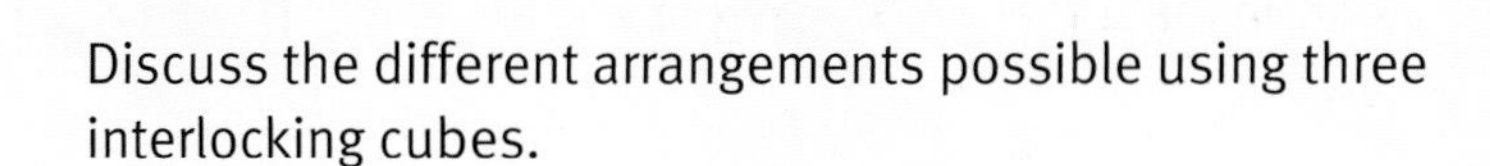

Discuss the different arrangements possible using three interlocking cubes.

How many different arrangements are possible?

Have you got them all? How can you be so sure?

Are these two shapes the same? Why? Why not?

The challenge

If appropriate, provide each pair with a copy of RS29. Ask the children to repeat the above activity, using the cubes to design houses with four rooms. Again, children work individually, then join with their partner to compare arrangements.

When they have made all the arrangements they can think of, children sort their shapes. They may do this by putting those with lengths of 4, 3, and 2 together, by combining those that are flat and those that 'stick out', or trying out other ideas. The purpose is to get children to look carefully at the construction of the shapes to see if they are really distinct or if they are duplicates in a different position.

If they count reflections as separate, children must make sure they have included reflections of all shapes. In the following example, if the children consider the two on the left to be different, they should also provide the matched pair to the one on the right. Encourage them to be consistent and logical.

The scrutiny of the shape will also encourage children to see if they have 'missed' any arrangements.

Drawing out using and applying

Children describe their shapes and how they sorted them. They explain how they know that they made all the possible arrangements of four cubes. They should be able to describe logical and systematic processes, starting with one arrangement and changing one cube at a time to make a new shape.

How are these two shapes the same? How are they different?

Have we found all the possible arrangements? How can we be sure?

What did you do to make sure that you found all the different arrangements?

Assessing using and applying

- Children can describe their shapes in general ways such as: "These are the same as those, but the other way round." Or: "These are flat shapes, but these can't be flat because they always stick up."
- Children can recognise reflections of the same shape.
- Children can arrange the cubes in a systematic way, keeping track of what they have tried before, until they know they have made all the possible arrangements.

Supporting the challenge

- Children spend additional time providing examples of shapes that have duplicates in a different position.
- Children use cubes of the same colour.

Extending the challenge

- Children find any lines of symmetry on each shape.
- Children find all the ways of arranging five cubes.

Challenge 26

Two cuts

Making 2D shapes and identifying some of their properties

Using and applying

Solving problems

Solve problems involving shape

Representing

Represent a puzzle using diagrams

Maths content

Understanding shape

- Draw and make polygons
- Classify polygons by identifying their properties

Key vocabulary

polygon, regular, irregular, pentagon, heptagon, rhombus, right-angled triangle, kite, hexagon, trapezium, irregular quadrilateral, isosceles triangle, octagon, parallelogram, oblong, equilateral triangle

Resources

For each child:

- Scrap paper squares
- Ruler, pencil
- RS30
- Scissors

For *Supporting the challenge*:

- Paper squares

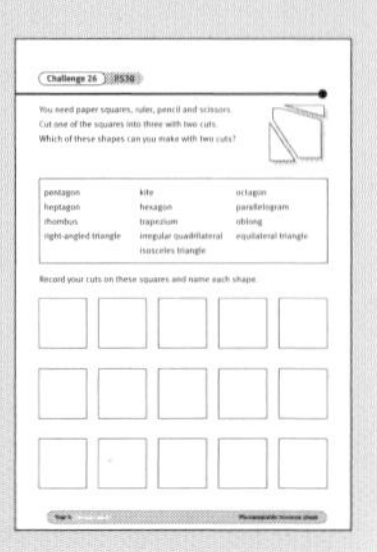

Which polygons can you make, using a square and two straight cuts?

Introducing the challenge

Give each child with the resources. Invite them to draw one straight line somewhere across their square and to cut along the line. Select some of these shapes to stick to the board.

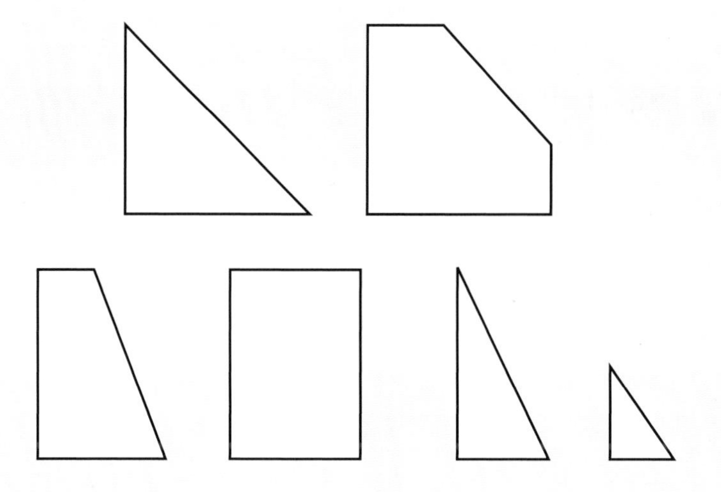

Ask children to sort them out into shapes that are the same in some way.

How many different kinds of triangle did we make? Has anyone got another kind?

Have we got any more kinds of rectangles? Some are wider and some are narrower.

Are there any other four-sided shapes? Any more quadrilaterals?

What is the most number of sides we have? What is that shape called?

The challenge

Give each child a copy of RS30. Set the children off on the challenge. Emphasise that they are only allowed to make two straight cuts for each shape. Demonstrate this if necessary.

Do you think you will be able to make each of these shapes?

Which shapes don't you think you will be able to make? Why?

Once children have completed the challenge, they compare and discuss their shapes with a partner.

Drawing out using and applying

Bring the class back together again and ask individual children to show and describe the shapes they made.

Which shapes were you able to make?

How did you make the parallelogram? Did anyone make it a different way? Is there another solution?

Which do you think is the most efficient solution? Why?

Assessing using and applying

- Children can choose to make cuts and see what shapes they arrive at.
- Children can choose a shape and search for ways to make it.
- Children can say why a shape is impossible. For example: "With one cut, the most sides you can make is five. With a second cut, you can only make a hexagon. So the octagon and heptagon are impossible."

Supporting the challenge

- Children continue to make shapes, using one cut.
- Limit the number of polygons you ask the children to try and make.

Extending the challenge

- Children make shapes, using three cuts.
- Children start with a triangle, hexagon or other shape of their choice.

Routes round school

Challenge 27

Using language of shape and position

Using and applying

Enquiring

Suggest a line of enquiry and the strategy needed to follow it; collect, organise and interpret selected information to find answers

Communicating

Report solutions to problems, giving explanations and reasoning orally and in writing, using diagrams and symbols

Maths content

Understanding shape

- Use the eight compass points to describe direction
- Use mathematical language to describe and identify position

Key vocabulary

position, route, direction, map, plan, compass, north, south, east, west, north-east, north-west, south-east, south-west

Resources

- Map of routes between two well-known places in the local area, written directions (made into an OHT or a IWB image), eight A4 sheets marked N, S, W, E, NE, SE, NW, SW, sticky tape

For each pair:

- RS31 (optional), small compasses, ruler, scrap paper and squared paper

How do you get from the main entrance to the school building to our classroom?

Introducing the challenge

Display the map and written directions between the two well-known places in the local area. Discuss the features of the map and the written directions with the children.

What makes this map easy to understand?

How do the written directions help?

Explain to the children how similar plans and written instructions can be used for shorter routes. Name two distinctive places in the school: for example, the main entrance to the school building and the door to the classroom.

Can anyone give me instructions that would guide someone from the main entrance to the school to our classroom?

Could anyone use these instructions to get to our classroom?

Could anyone describe a different route?

Discuss the eight compass points with the children. Point to north and stick the N sign on that wall. Do the same with S, E, W. Tell the children to face north and to point to the east. Then ask them to move their arms to point in the direction that is halfway between north and east. Remind the children that this is called north-east. Stick the NE notice on the appropriate wall. Do the same with the other directions.

Briefly discuss the route from the classroom door to the school entrance in relation to the compass points.

Point to roughly where the school entrance is. Which direction is that in?

If you leave the classroom and turn left, roughly what direction is that in?

The challenge

Introduce the challenge to the children, using RS31 if helpful. It is unlikely that a child could draw a sketch plan of the school accurately the first time. Encourage the children to draft some outline plans before trying to complete a more accurate version.

What features on your plan are important?

Is there another way to go between your two places?

Which is the best way? Why?

How could you improve your plan?

Once the children have modified their plan to create the best plan possible, they write instructions for their route so that someone with the plan could follow it. Most children are likely to use instructions such as 'left', 'right' and 'straight on', rather than compass directions.

Drawing out using and applying

Discuss the children's plans.

How are our plans similar?

What information do some plans have that others do not? Is this information necessary? Why? Why not?

Which direction is north on your plan? Can you tell from where you have drawn the classroom?

Children read out or describe some of their routes. Can the others follow them on their plans? Can they follow the instructions without the plans?

What do the best instructions have?

Is the most direct route always the best route?

Assessing using and applying

- Children can omit important pieces of information from their plans.
- Children can accurately record parts of the school with respect to their relative position.
- Children can accurately record parts of the school with respect to their relative distances.

Supporting the challenge

- Children choose two places within the classroom and describe a route from one to the other, avoiding obstacles.
- Tell the children the two places they are to draw their plan and write their instructions for.

Extending the challenge

- Children make sketch plans of their routes from home to school.
- Children include compass directions on their plans.

How big is an egg?

Choosing and using standard units and measuring instruments

Challenge 28

Using and applying

Solving problems

Solve problems involving measures

Representing

Represent a problem using statements or diagrams; use these to solve the problem

Maths content

Measuring

- Choose and use standard metric units and their abbreviations when estimating, measuring and recording
- Choose and use appropriate instruments to estimate, measure and record

Key vocabulary

length, width, weight, unit, instrument

Resources

For each group:

- RS32 (optional)
- A selection of small, medium and large eggs
- Modelling clay
- Measuring equipment: tape measure, calibrated jugs, displacement jug, water, callipers, scales and weights

For *Extending the challenge*:

- Construction material: card, glue, scissors

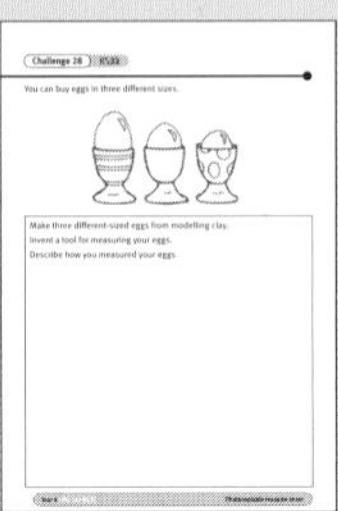

Can you invent a tool for measuring whether eggs are small, medium or large?

Introducing the challenge

Find out what the children know about buying eggs and the sizes they are sold in. Share their ideas about how they think the eggs are measured. Children may suggest a variety of ideas, from height and circumference to weight and possibly volume. Encourage and accept all these ideas.

Ensure that each group has access to all the resources and measuring equipment. If necessary, briefly remind the children of the purpose and usage of some of the equipment. Children then find out what they can about the sizes of eggs, using all the equipment available.

The challenge

In groups, children try to evolve a definitive description of a small, medium and large egg, in terms of some specific measurement or set of measurements. They may decide on length and width or circumference, weight, volume by displacement, or a combination of these. Children can use RS32 if helpful.

When they have done this, they invent a 'machine' or 'tool' for measuring eggs in order to grade them. Use mock eggs made from modelling clay for this part of the challenge.

How do you know that this is a large egg?

In what ways is it larger than this egg?

How can you measure it?

Drawing out using and applying

Share the results with the whole class. Compare the results from each group by having them report back on their work. Discuss what differences there are between groups in their definitions of small, medium and large eggs: for example, differences in types of measures used, accuracy of measures, sample of eggs used and estimations made.

What makes a small/medium/large egg?

How did your definition of what makes a small/medium/large egg influence the measuring tool you invented?

Children share and compare tools for measuring eggs. They may have invented a circumference gauge, a set of weighing guidelines, a displacement measure or other gadgets. You could ask a group to test one of each sized egg to find out if each tool grades them in the same way.

What are the benefits of Yolanda's group's device?

How might it be improved?

These two tools grade eggs in the same way, but this tool doesn't. Why do you think that is?

Which tool is the more accurate? What makes it so accurate?

Assessing using and applying

- Children can decide how to measure the eggs and what tools to use.
- Children can measure the eggs accurately enough in the way they have chosen and use appropriate units.
- Children can draw conclusions about how to grade eggs from the information they have found out and apply this to invent a way of grading eggs.

Supporting the challenge

- Some groups may have difficulty in deciding upon one particular way of measuring the different sizes of the eggs. Guide them towards any one method.
- Encourage groups to design a measuring tool that is based on their group's decision about how egg sizes are measured.

Extending the challenge

- Children investigate the sizes and shapes of other birds' eggs: quail, duck, goose.
- Children design a box for holding eggs.

Slow helicopter

Choosing and using units of time and using knowledge of shapes

Challenge 29

Using and applying

Solving problems

Solve problems involving measures, including time

Representing

Represent a problem using number sentences, statements or diagrams; use these to solve the problem; present and interpret the solution in the context of the problem

Maths content

Measuring

- Choose and use standard metric units and their abbreviations when estimating, measuring and recording length; use decimal notation to record measurements
- Read time to the nearest minute; choose units of time to measure and calculate time intervals

Key vocabulary

length, metre (m), time, minutes, seconds, shape

Resources

For each pair:

- A4 scrap paper, scissors, ruler, stapler, hole punch, sticky tape, glue, elastic bands, paper clips; metre rule, timer or stopwatch

For *Supporting the challenge*:

- A4 paper helicopters, RS33

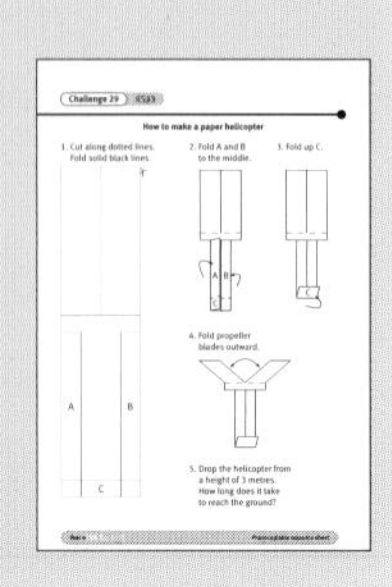

Challenge 29 RS33

How to make a paper helicopter

1. Cut along dotted lines. Fold solid black lines.

2. Fold A and B to the middle.

3. Fold up C.

4. Fold propeller blades outward.

5. Drop the helicopter from a height of 3 metres. How long does it take to reach the ground?

For *Extending the challenge*:

- Tissue paper, balloons, string

Can you design a helicopter that falls as slowly as possible?

Introducing the challenge

This challenge is best undertaken when the children are studying work in science on flight, possibly after looking at 'helicopter' seedlings.

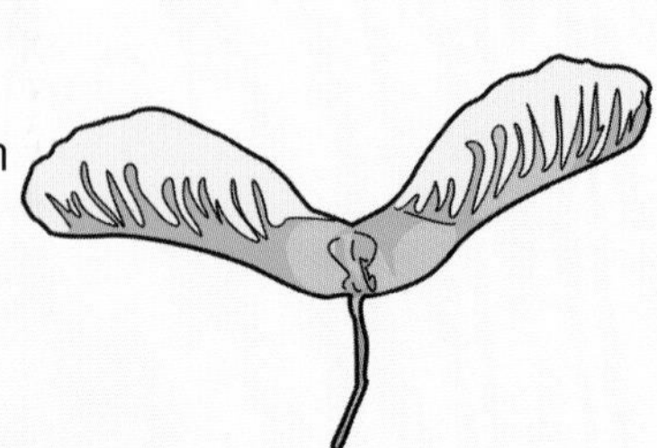

Introduce the challenge to the children: they need to make a paper helicopter that takes the longest time to fall to the ground from a height of 3 metres, using a sheet of A4 paper (see RS33). Briefly discuss the challenge with the children and gather together some initial ideas from the class.

How long should the propeller blade be? How long should the tail be?

Will you need a paper clip on the bottom of the tail? Why? Why not?

Explain to the children how they should discuss, plan, make and try different designs in pairs until they feel they have made the best helicopter possible.

The challenge

Provide a dropping point either in a clear space in the classroom, reached by standing on a chair, a suitable place in the gym hall or an accessible sheltered place in the playground. Children can use whatever design they like, but they can only use one sheet of A4 paper for each helicopter (so the dimensions of the helicopter are restricted).

Children need to try several designs to get a good flier, then modify it to fly as slowly as possible. Their timing should be as accurate as possible, as the flight will be short. They may want to establish rules about dropping the helicopter and when it is deemed to have landed. Children record their progress, including modifications and reasons for them, and draw and describe their final design.

What was your first helicopter like?

How did you improve on it? Why did you make that modification?

How is your final design different from your first design? What was the main alteration you made?

How did you measure the time it took your helicopter to land?

Drawing out using and applying

When the children have produced their best possible results and recorded them, organise a demonstration with the whole class, with a timekeeper. The timings of the demonstration may vary from the children's own tests; ask them to suggest reasons for this. Each pair can also describe how they developed their final version, what adaptations they made, and why.

How did your helicopter change through the various versions you had?

What makes your final version the best helicopter?

What design features do the most effective helicopters all have?

How could some of your helicopters be improved?

Assessing using and applying

- Children can find a way to begin the problem and use their knowledge of 'helicopter' seeds to make a starting model.
- Children can set up a fair test for their model and can find a method of measuring the flight time accurately.
- Children can modify and improve their designs in the light of their tests, identify what needs improving and explain why.

Supporting the challenge

- Provide the children with some simple A4 paper helicopters to get them started (see RS33).
- Discuss with the children how making the helicopter rotate will slow down its fall to the ground.

Extending the challenge

- Children investigate paper aeroplanes and develop the fastest plane to travel over the longest distance. They find ways of measuring both the distance and the time.
- Children build model air balloons and parachutes to make something that will be airborne as long as possible.

Making weights

Challenge 30

Exploring and using standard units of weight

Using and applying

Reasoning

Identify and use patterns, relationships and properties of numbers

Communicating

Report solutions to problems, giving explanations and reasoning orally and in writing, using diagrams and symbols

Maths content

Knowing and using number facts

- Use knowledge of addition and subtraction facts and place value to derive sums and differences of pairs of multiples of 10

Measuring

- Choose and use standard metric units and their abbreviations when estimating, measuring and recording weight

Key vocabulary

weigh, weight, balance, kilogram (kg), gram (g), light/lighter/lightest, heavy/heavier/heaviest

Resources

For each pair:

- RS34, modelling clay
- 100 g standard weight, balance, objects in the classroom to weigh
- Small pieces of card

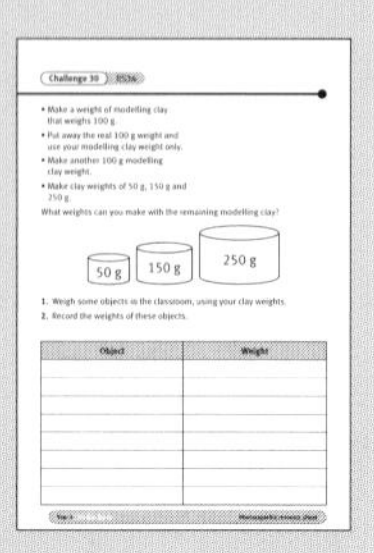

Can you weigh other things with 100 g as your starter weight?

Introducing the challenge

Show the children a balance and remind them how it works. Using the 100 g weight on one side of the balance, place weights on the other to show equal and non-equal weights.

Is the weight on each side of the balance the same? How do you know? Which side is lighter? Which side is heavier?

Give each pair of children a copy of RS34. Introduce the activity and ask children how they think they might make different weights with modelling clay, using just 100 g. Make sure children understand the relationship between different units of weight such as 25 g, 50 g, 250 g, 500 g and 1 kg. Children then work on the problem in pairs.

How many grams are there in 1 kilogram? What about half a kilogram?

What fraction of a kilogram is 250 grams? What about 100 grams?

The challenge

Give each pair modelling clay, a balance and a 100 g standard weight as well as a variety of classroom objects to weigh. One difficulty with the problem is resisting the temptation to remove some modelling clay from a made-up weight in order to balance a new piece of modelling clay. Observe children as they do the activity to see how they solve this problem, and also the problem of halving the modelling clay to make two equal weights.

How will you make 150 grams?

How could you make a 75 gram weight?

How many clay weights could balance the large clay weight? How do you know?

What is the weight of the pair of scissors? What about the exercise book?

Drawing out using and applying

Children show one or two of the objects they weighed, using their clay weights. Provide them with pieces of card to label the weight of the object and build up a class display of some of the objects weighed.

Different pairs of children compare their modelling clay weights to make sure they weigh the same.

How could we check to see if these two 150 gram weights weigh the same?

Let's see if the 50 gram weight and the 100 gram weight that Chaim and Leo made weighs the same as Freda and Angelina's 150 gram weight.

Discuss with the class some of the difficulties children experienced making their clay weights.

Once you had made your 100 g weight, how did you use it to make the other weights?

Which weights were easy to make? Which were more difficult? Why was that?

Add some of the modelling clay weights to the display.

Assessing using and applying

- Children can use a 100 g modelling clay weight as the basis for making other weights.
- Children can solve problems logically: for example, divide 100 g of modelling clay into two equal halves to make two 50 g weights and then combine these with other weights to make different combinations.
- Children can explain how they made each weight and how they could make different weights such as 125 g or 75 g weights.

Supporting the challenge

- Make the 100 g and 50 g modelling clay weights for the children. This should help them make their 150 g and 250 g weights.
- Provide extra assistance when the children are making their 50 g, 150 g and 250 g modelling clay weights.

Extending the challenge

- Children make modelling clay weights as small as they can and weigh very small objects.
- Children find out if ten 100 g clay weights weigh the same as a 1 kg weight and explore reasons for any discrepancies.

Cover it

Challenge 31

Calculating area

Using and applying

Reasoning

Identify and use patterns, relationships and properties of shapes; investigate a statement and test it with examples

Communicating

Report solutions to puzzles, giving explanations and reasoning orally and in writing, using diagrams and symbols

Maths content

Understanding shape

- Visualise 3D objects from 2D drawings

Measuring

- Find the area of rectilinear shapes by counting squares

Key vocabulary

area, length, width, depth, square, pentominoes, grid

Resources

For each child:

- Lots of interlocking cubes

For each pair:

- RS35
- Squared paper
- Ruler

Which shapes are good for covering a surface?

Introducing the challenge

Ask each child to take four interlocking cubes and join them together to make a shape that is one cube deep. Compare the different shapes made and ask if anyone can make one that is different from those already made.

Describe one of your shapes.

Who made a different shape to this?

Are there any more we can make, using four interlocking cubes?

How many are there?

What about these two shapes? Are they the same or different? How are they similar?

- **There are five different four-cube shape possibilities if you do not count shapes that can be made by turning one over.**

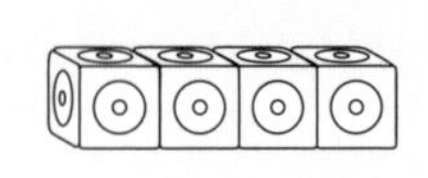 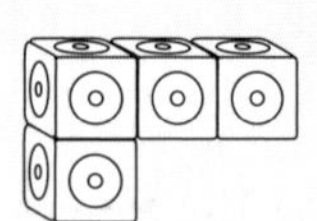 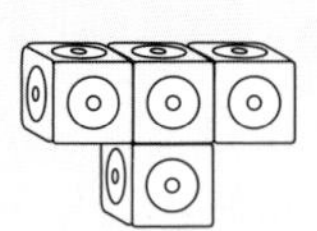 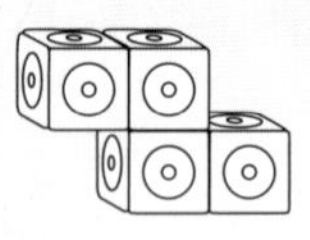 

The shapes below are treated as identical.

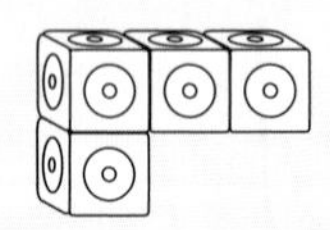 

The challenge

Arrange the children into pairs and give each pair a copy of RS35, squared paper and a ruler. Talk through the first part of the challenge.

Look at the five different four-cube shapes you have just made. Which shape, when duplicated four times, will fit together to make a 4 by 4 shape?

Will any of the other four-cube shapes fit together to make a 4 by 4 shape when duplicated four times?

Children work with their partner to record their findings on 4 by 4 grids on squared paper.

Next, tell the children to do the same, but this time using five-cube shapes duplicated five times and fitting them together to make a 5 by 5 square.

- **There are 12 different five-cube shapes possible.**

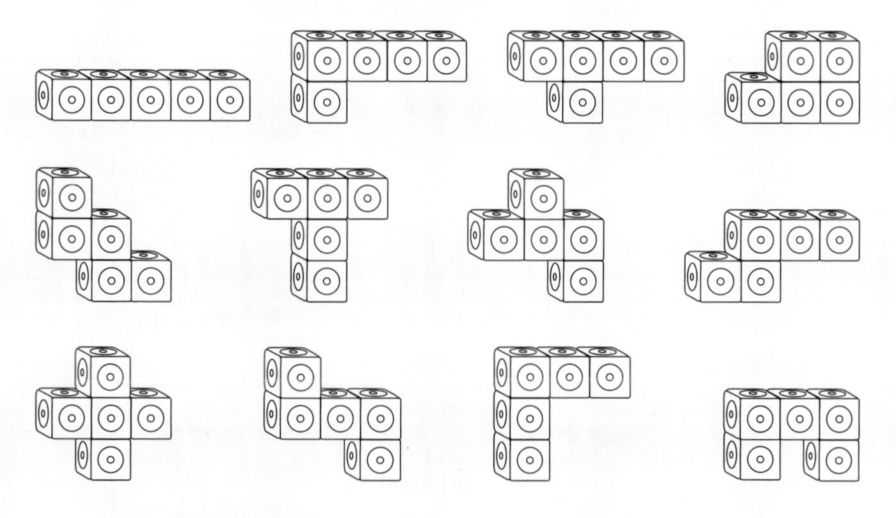

That is, if you do not count shapes that can be made by turning one over.

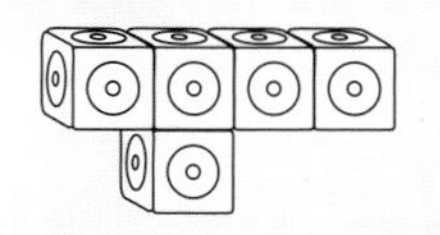

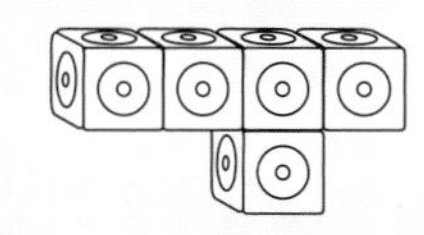

Drawing out using and applying

Children share their results for the four-cube shapes before moving on to the five-cube shapes.

Which shapes fitted together to make a 4 by 4 square? A 5 by 5 square?

How can you be sure that some shapes did not fit together?

Why couldn't you use this shape to make a 4 by 4 square? A 5 by 5 square?

How many of the four-cube shapes made a 4 by 4 square? How many of the five-cube shapes made a 5 by 5 square?

Assessing using and applying

- Children can make shapes and use trial and improvement to see if they work.
- Children can decide which shapes do not work by exhausting the possible arrangements.
- Children can present a logical argument as to why some shapes do not work. For example: "Two of this shape can be fitted together to make five squares wide, but filling the remaining three-wide space is impossible."

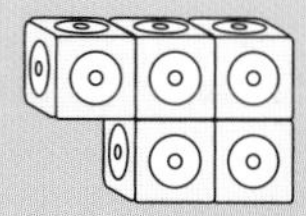

Supporting the challenge

- The children work on the first part of the challenge: the four-cube shapes.
- Rather than recording the shapes on squared paper, children show shapes that go together to make a 4 by 4 square using colour to differentiate the four four-cube shapes. For example:

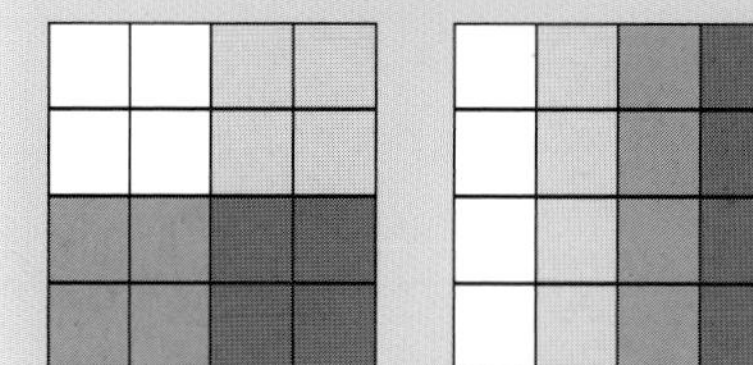

Extending the challenge

- Children make shapes with six cubes and fit them together to make a 6 by 6 square.
- Which four-cube shapes will fit together to make a 4 by 4 by 4 cube?

Bags of fun

Challenge 32

Exploring the relationship between area and perimeter

Using and applying

Solving problems

Solve problems involving measures

Representing

Represent a problem using statements or diagrams; use these to solve the problem

Can you make two different bags with edges measuring 1 metre?

Introducing the challenge

Show the children the collection of flat paper bags. Talk about the kind of constraints that manufacturers have in order to make things economically. Discuss how the bags are constructed.

Explain that the challenge is to make two bags with an economical constraint. Each bag has to have a perimeter of 1 metre.

Remind the children that the term 'perimeter' means the distance around the boundary of a shape. In the case of the bags, it is the distance around the edge of the bag when it is laid out flat.

Explain that one way of finding the perimeter of a rectangular shape is to measure the length of two perpendicular sides, add them together and double their answer.

The challenge

Arrange the children into pairs and give them a 1-metre strip of paper or tape (or ask them to cut off 1 metre of tape from a roll) and a copy of RS36 if helpful. Explain the challenge to the children.

To manufacture them economically, eco bags have to have a perimeter of 1 metre. Can you make two different bags with edges measuring 1 metre?

Ensure that each pair of children have rulers and tape measures, a variety of paper in different sizes, glue, sticky tape, scissors and any other appropriate construction material that may be helpful.

Encourage children to plan what their bag is going to be used for as well as what it will look like.

Maths content

Understanding shape

- Visualise 3D objects from 2D drawings and make nets

Measuring

- Draw, measure and calculate perimeters; find the area of rectilinear shapes

Key vocabulary

perimeter, area, centimetre (cm), shape

Resources

For each pair:

- Collection of flat paper bags (at least one)
- 1-metre strip of paper or tape
- RS36 (optional)
- Ruler and tape measures
- Large sheets of paper, glue, sticky tape, scissors
- Other appropriate construction materials

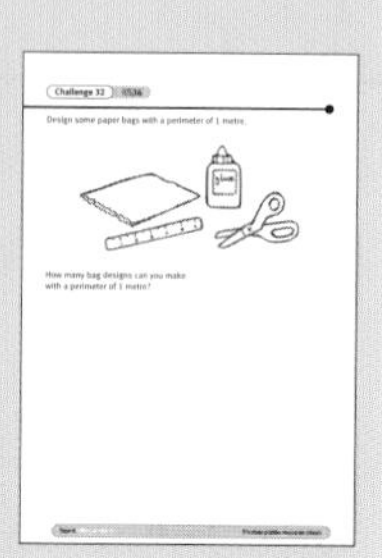

Drawing out using and applying

Ask individual pairs of children to explain to the rest of the class how they decided on the shape and size of their bags.

Can you show us that the perimeters of both bags measure 1 metre?

How can you measure round corners? Where do you start measuring from?

Compare the various bags that the children have made and make a display of them ranging from tall and thin to squarish to short and squat. Invite one or two children to check that the perimeter of each bag is about 1 metre.

Do you think both these bags have the same perimeter?

Which bag do you think would contain more?

Work with the class to draw up a list of criteria for what they think makes a bag fit for its purpose.

What things did you need to think about when designing your bag?

Which of these were influenced by the purpose of your bag?

How would you rate your bag against these criteria?

How could you improve your bag?

Assessing using and applying

- Children can determine a shape with an equivalent perimeter by trial-and-improvement methods.
- Children can decide which shape they want their bag to be and determine an appropriate size.
- Children can coordinate the shape of their bags with its purpose and design it accordingly.

Supporting the challenge

- Encourage children to start with the 1-metre strip of tape and to outline different shapes, then decide how to make the bags.
- Suggest children carefully cut a commercial bag to reveal the net of the bag. Discuss with them how this can help them make their own bag.

Extending the challenge

- Children design a box for a particular purpose: for example, to hold an egg securely.
- Children work out how to make a paper carrier bag with a gusset.

Sports day

Challenge 33

Solving problems involving time

Using and applying

Enquiring

Suggest a line of enquiry and the strategy needed to follow it; collect, organise and interpret selected information to find answers

Communicating

Report solutions to problems, giving explanations and reasoning orally and in writing, using diagrams and symbols

Maths content

Knowing and using number facts

- Use knowledge of rounding, number operations and inverses to estimate and check calculations

Measuring

- Read time to the nearest minute; use am, pm and 12-hour clock notation; choose units of time to measure and calculate time intervals

Key vocabulary

measure, time, interval, estimate

Resources

For each group:

- RS37 (optional), scrap paper, large sheet of paper, marker

For *Supporting the challenge*:

- Paper strips in different lengths
- Photos, programmes, trophies and other memorabilia from past sports days

Challenge 33 RS37

National School Sports Day Competition!
Planning the perfect sports day

Competition rules

Competitors must include details of the following:

- the length of the day
- the sporting events
- the timing of the events
- the number of events
- winning rules for the judges
- catering arrangements
- the awards ceremony
- events for parents

Competitors will be expected to present their plans to the other competitors.

A vote will be taken to find the winner.

What would your perfect sports day be like?

Introducing the challenge

Discuss sports days and what the children like most about them.

What are your favourite events?

What don't you like about sports day?

Tell the children that, as a group, they are going to plan their perfect sports day.

On the board, brainstorm what they will have to think about. Focus not only on the events but also on the allocation of time for each event and other occasions that are likely to occur throughout the day: for example, lunch, the award ceremony, a parents race.

What things will you need to think about?

How long will it last?

How many events will you have?

Will the events be the same for each year group?

How will you fit them all in?

The challenge

Arrange the children into groups and encourage them to discuss and share ideas before getting down to the detail of planning the day. Help the children think about further questions that they need to answer. You may wish to provide each group with a copy of RS37 to jot down their initial ideas.

How will you decide on reasonable lengths of time to allow for each event?

Before each group begins to write out their plan for the day, provide them with a large sheet of paper and a marker. Explain that they are to use the sheet to write out their plan for the sports day. Tell the groups that they will use this as a crib when it is their turn to convince the rest of the class that their sports day would be a really fun day.

Drawing out using and applying

Groups present their plans.

Machui's group, tell us about your plan. What would your perfect sports day be like?

What do you think are the best features of Machui's group plan?

Which aspects do you think could be improved? How?

Is there anything that this group has forgotten about?

Once each group has presented their plan to the class, have a class vote for the perfect sports day.

Explain to the children that they are to think carefully about each of the plans and decide which one they think is the perfect plan. Tell them that they can only vote once and that they are not allowed to vote for their own group's plan.

Referring to each group in turn, ask the children to raise their hands if they think this is the perfect plan.

Assessing using and applying

- Children can plan a single event without having to consider people's interests or time constraints.
- Children can take some constraints into account: for example, that two running races cannot happen at the same time.
- Children can plan a time line for the events and fit in events to maximise the use of time.

Supporting the challenge

- Use strips of paper to represent different events: the length of the paper should be in proportion to the time the events will last. Help the children arrange these either sequentially or in parallel to plan their event.
- If available, provide photos, programmes, trophies and other memorabilia from past sports days.

Extending the challenge

- Children plan a refreshments stall for the sports day.
- Children plan a class outing.

What's your name?

Challenge 34

Gathering data and displaying it appropriately

Using and applying

Representing

Represent problem using number sentences, statements or diagrams; use these to solve the problem; present and interpret the solution in the context of the problem

Enquiring

Suggest a line of enquiry and the strategy needed to follow it; collect, organise and interpret selected information to find answers

Maths content

Handling data

- Answer a question by identifying what data to collect
- Organise, present, analyse and interpret the data in tables, diagrams and tally charts

Key vocabulary

data, information, table, collect, organise, present, analyse, interpret

Resources

- RS38 (optional, for each group)
- Class registers or name lists

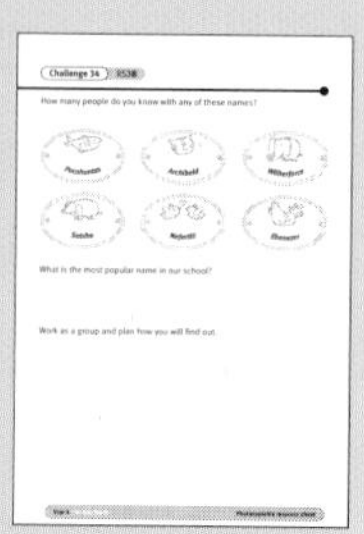

What is the most popular name in our school?

Introducing the challenge

Pose the following questions to the children:

Does anyone in the class have the same name as anyone else?

What about the same name as somebody's brother or sister?

Discuss the questions with the children and the fact that names vary in popularity over the years.

Pose the following question to the children:

What is the most popular name in our school?

Discuss the question with the class. Ask the children what information they think they would need to gather to answer the question. Children justify their answers.

What do you think might be the most popular name in our school?

Why do you think this?

How might we find out for sure?

Discuss with the children how they might find out the answer to the question and how they might go about recording their results.

How might you go about recording the information you find out?

What are the different ways you could present your findings?

Which might be the most appropriate? Why?

The challenge

Set the children off to work in groups, recording their findings on RS38. Point out that you are not going to allow them to visit other classrooms! Tell the groups that they can only have access to the registers or name lists when they have come up with a plan for organising the way they are going to work.

Encourage the children to think about the best way of presenting their findings.

How are you recording your findings? Is this the best way to help you interpret your results?

How else might you record your results?

Drawing out using and applying

Gather the groups together to share their findings. Discuss with the class the way different groups recorded their findings, and which were the easiest to interpret and why.

What is the most common name in our school?

Did any group come up with a different name?

How did you organise yourselves to work?

Who organised themselves differently?

Which method of organisation do you think is the most efficient? Why?

How did you present your results? Did this make them easy to read and interpret?

How could you have presented your findings differently? Would this have been better? Why?

Assessing using and applying

- Children can organise themselves effectively and do not duplicate work.
- Children can take some form of leadership and organise the group to undertake a specific task.
- Children can discuss various strategies and agree on the best approach.

Supporting the challenge

- Provide the children with photocopies of one or two class lists.
- Assist the children in organising and presenting their results.

Extending the challenge

- Children investigate popular names of their parents or grandparents age.
- What is the most popular surname in your school? What is its country of origin?

Books in school

Making estimations and gathering data to confirm estimates

Challenge 35

Using and applying

Enquiring

Suggest a line of enquiry and the strategy needed to follow it; collect, organise and interpret selected information to find answers

Reasoning

Identify and use patterns, relationships and properties of numbers; investigate a statement involving numbers and test it with examples

Maths content

Handling data

- Answer a question by identifying what data to collect
- Organise, present, analyse and interpret the data

Key vocabulary

data, information, collect, organise, present, analyse, interpret, estimate, mean, mode

Resources

- Individual whiteboard and pen (for each child)
- RS39 (for each group)

How many books do you think there are in our school?

Introducing the challenge

Ask the children to jot down on their whiteboards how many books they think are in the classroom. Decide with the children whether this is to include exercise books and books the children may have made or only commercially produced books.

Ask for some of the estimates and write them up on the board.

Dasha, what is your estimate?

Who has an estimate greater than this? Less than this?

How big is the difference between the greatest and the smallest estimates?

How accurate do your think these estimates are?

How accurate do you think an estimate should be to be acceptable?

The challenge

Provide each group with a copy of RS39 and discuss with the class how they might go about estimating the total number of books in the school. Explain that an exact count is not required, only a fairly accurate estimate.

Give the children about 10 minutes to share their ideas and suggest ways of gathering the data.

Invite groups to share their ideas with the rest of the class.

Ask the children, back in groups, to firm up their plans. Tell them that they need to have their plans approved by you before setting off on the task. Once you have approved each group's plan, allow them sufficient time to go about implementing it.

Once the groups have finished, they think about what else they could estimate the numbers of.

Drawing out using and applying

Gather together the various estimates that the groups arrived at. Invite individual groups to share their methods of data collection.

How many books do you think there are in our school?

How did you arrive at that estimate?

Do you think it is a fairly accurate estimate? What makes you so sure?

Who has a similar estimate to this?

Who has a vastly different estimate to this?

Why are they so different? Which do you think is the more accurate? Why do you think that?

How big is the range of estimates? What is the difference between the greatest and the smallest?

Work with the children to calculate the mean average of the estimates – that is, the total of all the estimates divided by the number of estimates. If appropriate, also help them identify the mode – that is, the estimate that occurs most often.

Is there any difference between the mean and the mode?

Which do you think is the better estimate, the mean or the mode? Can you explain why?

Finally, invite groups to offer some of their suggestions as to other things they could estimate the numbers of.

How might you go about collecting the data for this?

What would be the same as the way you went about finding out how many books there are in the school? What would be different?

Assessing using and applying

- Children can use counting methods in the classroom to estimate numbers of books.
- Children can share out the task: for example, a pair working on fiction, another on non-fiction.
- Children can use calculations to estimate: for example, counting the number of books in a length of shelving and multiplying by the number of shelves.

Supporting the challenge

- Focus the children on estimating, for example, the number of books in just part of the classroom or on a particular shelf. Support them in thinking about how they could check their estimate without actually counting.
- Children focus on the second challenge on RS39 and estimate how many books they think there are in the classroom.

Extending the challenge

- Children estimate other collections: for example, pencils, chairs, interlocking cubes, and so on.
- Children estimate events in time: for example, how long they spend each week/term/year on maths or how many hours a year they spend in school.

How far from home?

Challenge 36

Answering and asking related questions, and collecting and presenting data

Using and applying

Enquiring

Suggest a line of enquiry and the strategy needed to follow it; collect, organise and interpret selected information to find the answer

Communicating

Report solutions to problems, giving explanations and reasoning orally and in writing, using diagrams and symbols

Maths content

Handling data

- Answer a question by identifying what data to collect
- Organise, present, analyse and interpret the data in tables, diagrams, tally charts, pictograms and bar charts

Key vocabulary

data, information, table, collect, organise, present, analyse, interpret

Resources

For each group:

- RS40
- Local maps, street map, A–Z of the area (all with a scale)
- Squared and/or graph paper
- Ruler
- Coloured pencils

For Extending the challenge:

- ICT data handling package

Challenge 36 RS40

How far away from the school is your home?

Work as a group and organise a survey of how far everyone in the class lives from the school.

Find a way of answering these questions:

1. Who lives less than a kilometre, in a straight line, from the school?
2. Who lives more than a kilometre, in a straight line, from the school?
3. Who lives furthest from the school?
4. Who lives nearest to the school?
5. How far is your actual journey to school?
6. How much longer is it than a straight line?
7. Imagine you need to reduce the size of the class by 10 children. You will do it by removing children who live far away. What rules will you decide upon? Who will you have to remove from the class?

Ask more questions of your own. Be prepared to find out the answers!

How far away from the school is your home?

Introducing the challenge

Pose the following question to the children:

How far away from the school is your home?

Discuss the question with the class. Ask the children for ideas about what information to collect and how to collect it. Be prepared to discuss with the class the distance between school and home as being measured in length or time.

What information would we need to collect to answer this question?

How would we go about doing it?

Children suggest further related questions to ask and how to collect the information. If the children do not ask it, pose the following question to the children:

How far does everyone in our class live from the school?

Again, discuss the question with the children, including ideas about what information to collect and how to collect it.

How would you go about collecting this information?

The challenge

Provide each group with a copy of RS40 and the other resources. Draw the children's attention to the list of questions. Explain to the children how they need to consider Ehow they are going to collect and present the results of their survey in order to find answers to these and their own questions.

How will you display the data?

What sort of graph might you try?

Encourage the children to organise their work.

Would it help to put things in order?

Is there any information you haven't got yet?

Is there another way of showing the data apart from a block graph?

Children are likely to have good ideas about how to tackle this investigation, but they will need help in organising and structuring it. Encourage them to think about this before you give the solutions.

Drawing out using and applying

Ask each group to do a short, formal presentation of their questions and their findings. This should include a written summary of their work and any graphs, diagrams and maps they have produced.

Invite children from other groups to ask questions.

Assessing using and applying

- Children can ask questions relevant to the problem. For example: "If someone lives at the top of a block of flats, are they the same distance from the school as someone who lives at the bottom?"
- Children can decide which information to collect and how to collect it: for example, using the class register for children's addresses and marking them on a street map, using the children's initials.
- Children can present their work in maps, graphs or diagrams and explain it clearly: for example, showing on a map a boundary line of one kilometre from the school, with numbers of children living inside and outside the border clearly marked, and explaining what the map shows.

Supporting the challenge

- Work first with the children on gathering information in terms of the time it takes people to get to school.
- Assist the children in organising and presenting their results.

Extending the challenge

- Ask the children to present their data using an ICT package.
- Children investigate times it takes to get to school. Does the person who lives furthest away take the longest to get to school?